NEUROMETHODS

Series Editor
Wolfgang Walz
University of Saskatchewan
Saskatoon, Canada

For further volumes:
http://www.springer.com/series/7657

Neuromethods publishes cutting-edge methods and protocols in all areas of neuroscience as well as translational neurological and mental research. Each volume in the series offers tested laboratory protocols, step-by-step methods for reproducible lab experiments and addresses methodological controversies and pitfalls in order to aid neuroscientists in experimentation. *Neuromethods* focuses on traditional and emerging topics with wide-ranging implications to brain function, such as electrophysiology, neuroimaging, behavioral analysis, genomics, neurodegeneration, translational research and clinical trials. *Neuromethods* provides investigators and trainees with highly useful compendiums of key strategies and approaches for successful research in animal and human brain function including translational "bench to bedside" approaches to mental and neurological diseases.

Light Sheet Microscopy

Edited by

Ludovico Silvestri

Department of Physics and Astronomy, University of Florence, Sesto Fiorentino, FI, Italy; European Laboratory for Non-linear Spectroscopy (LENS), Sesto Fiorentino, FI, Italy

Humana Press

Editor
Ludovico Silvestri
Department of Physics and Astronomy
University of Florence
Sesto Fiorentino, FI, Italy

European Laboratory for Non-linear Spectroscopy (LENS)
Sesto Fiorentino, FI, Italy

ISSN 0893-2336 ISSN 1940-6045 (electronic)
Neuromethods
ISBN 978-1-0716-5099-8 ISBN 978-1-0716-5100-1 (eBook)
https://doi.org/10.1007/978-1-0716-5100-1

This Humana imprint is published by the registered company Springer Science+Business Media, LLC, part of Springer Nature.
The registered company address is: 1 New York Plaza, New York, NY 10004, U.S.A.

Preface to the Series

Experimental life sciences have two basic foundations: concepts and tools. The Neuromethods series focuses on the tools and techniques unique to the investigation of the nervous system and excitable cells. It will not, however, shortchange the concept side of things as care has been taken to integrate these tools within the context of the concepts and questions under investigation. In this way, the series is unique in that it not only collects protocols but also includes theoretical background information and critiques which led to the methods and their development. Thus it gives the reader a better understanding of the origin of the techniques and their potential future development. The Neuromethods publishing program strikes a balance between recent and exciting developments like those concerning new animal models of disease, imaging, in vivo methods, and more established techniques, including, for example, immunocytochemistry and electrophysiological technologies. New trainees in neurosciences still need a sound footing in these older methods in order to apply a critical approach to their results.

Under the guidance of its founders, Alan Boulton and Glen Baker, the Neuromethods series has been a success since its first volume published through Humana Press in 1985. The series continues to flourish through many changes over the years. It is now published under the umbrella of Springer Protocols. While methods involving brain research have changed a lot since the series started, the publishing environment and technology have changed even more radically. Neuromethods has the distinct layout and style of the Springer Protocols program, designed specifically for readability and ease of reference in a laboratory setting.

The careful application of methods is potentially the most important step in the process of scientific inquiry. In the past, new methodologies led the way in developing new disciplines in the biological and medical sciences. For example, Physiology emerged out of Anatomy in the nineteenth century by harnessing new methods based on the newly discovered phenomenon of electricity. Nowadays, the relationships between disciplines and methods are more complex. Methods are now widely shared between disciplines and research areas. New developments in electronic publishing make it possible for scientists that encounter new methods to quickly find sources of information electronically. The design of individual volumes and chapters in this series takes this new access technology into account. Springer Protocols makes it possible to download single protocols separately. In addition, Springer makes its print-on-demand technology available globally. A print copy can therefore be acquired quickly and for a competitive price anywhere in the world.

Saskatoon, SK, Canada *Wolfgang Walz*

Preface

Light-Sheet Fluorescence Microscopy: A Game Changer

In a way, the history of neuroscience is the history of its methods. R. Yuste

Introduction

The brain is probably the most complex organ of our body. From a functional point of view, it collects and organizes an extremely large amount of sensory information, coordinates complex movements, and is capable of learning new behavioral schemes and of performing abstract reasoning. The neuroscience community agrees on the fact that these unique features are enabled by the extraordinary complex organization of the brain itself. Indeed, even if we limit ourselves to the relatively simple brain of a mouse, we have to deal with a complicated network of about 10^8 neurons, connected by about 10^{11} synapses [1]. In addition, neurons differ from each other from a morphological, chemical, and electrophysiological perspective. Different classifications have identified hundreds of cell types in the brain, based either on transcriptomics [2], firing patterns [3], morphology [4], or by the combination of multiple criteria [5–7].

Historically, neuroscience research has focused on the detailed investigation of the properties of single neurons, starting from the morphological analysis carried out by Santiago Ramón y Cajal exploiting single-cell staining with Golgi's black reaction [8]. We can insert in this reductionist framework also the foundational work on single-cell electrophysiology from Hodgkin, Huxley, and Katz at the middle of the last century [9, 10], and the latest developments in single-cell transcriptomics [11]. This large amount of research has provided us a solid ground to understand how the brain works at single-neuron level, but limited insights on the large-scale mechanisms that orchestrates the activity of all these cells. This partial view was not due to a conceptual prejudice, but rather technical limitations. Indeed, the interconnected nature of the brain has been acknowledged since the dawn of neuroscience [12]. However, available techniques to study morphological, physiological, and biochemical features of single neurons required to physically dissect the brain, limiting the possibility of observing large-scale organization principles.

In contrast to this reductionist approach, the advent neuroimaging methods in the last part of twentieth century brought the ability of measuring the orchestrated activity of the entire brain [13]. The advent of methods like functional Magnetic Resonance Imaging (fMRI) enabled the observation of brain-wide activity patterns in awake human subjects performing specific cognitive tasks [14, 15]. Other techniques, like Diffusion Tensor Imaging (DTI), allowed anatomical reconstruction of white-matter connectivity across the entire brain [16]. These technological advancements eventually resulted in a paradigm shift at the theoretical level. Indeed, it is hard not to link the rise of network neuroscience [17] with the possibility of measuring structure and function on a brain-wide scale.

Thus, as summarized by Rafael Yuste in the phrase used as epigraph of this chapter, "In a way, the history of neuroscience is the history of its methods" [18]. The techniques that we use in the lab determine the types of experiments we can perform and which data we collect and analyze, ultimately shaping our view of the brain. The reductionist and network approaches described above are the result of diverse technological developments. Unfortunately, linking the insights we obtained in cellular and molecular neuroscience with brain-wide neuroimaging studies is proving to be a formidable challenge. Indeed, the resolution achieved with the latter methods is insufficient to identify single cells. On the other hand, typical cellular neuroscience studies are limited to few cells in a very limited brain volume, losing the inherent complexity of the brain.

The logical conclusion is that we need a methodology capable of analyzing neuronal structure and function *with cellular resolution* and *on a brain-wide scale.* Such a technique will drive a new paradigm shift in neuroscience, bridging the gap between the reductionist analysis of single neurons and global analysis of the brain as a whole. Given the title of the book, it is not difficult to imagine that I have a candidate in mind: light-sheet fluorescence microscopy (LSFM). Indeed, this method has demonstrated the potential to analyze the entire mammalian brain with single-cell resolution and molecular specificity [19, 20], and has been used to study the activity of all the neurons (one by one) across the entire brain in simple vertebrates like Zebrafish [21].

Basic Features of Light-Sheet Fluorescence Microscopy

The hallmark of LSFM is uncoupling of excitation and detection optical paths, differentiating it from other fluorescence-based microscopies, like confocal [22] or two-photon [23]. Excitation is modeled as a thin sheet of light, which is coincident with the focal plane of detection optics. This selective illumination scheme[1] enables optical sectioning—i.e., 3D resolution—with a wide-field detection scheme—i.e., collecting images with a camera rather than performing a point scanning. LSFM can thus perform 3D reconstructions with a speed several orders of magnitude faster than confocal or two-photon microscopy [24], and is currently limited by the frame rate of available scientific cameras and the ability of computers to stream digital images to permanent memory. As an example, an entire mouse brain can be reconstructed with micron resolution in few hours with a light-sheet microscope [25].

The uncoupling between excitation and detection is evident also in the optical resolution of an LSFM system. Radial resolution—i.e., resolution in the illumination plane, perpendicular to detection axis—depends on the numerical aperture (NA) of the detection objective, following the classical Abbe's formula:

$$d_{\text{radial}} = \frac{0.61\ \lambda_{\text{f}}}{\text{NA}_{\text{det}}} \tag{1}$$

with λ_{f} being the central wavelength of fluorescence emission. Axial resolution—i.e., resolution along the detection axis—equals the thickness of the light sheet and thus depends on the illumination optics only. In the typical case of illumination with a Gaussian laser beam, we can approximate this thickness with the beam waist w_0, which is related to the illumination NA by the standard Gaussian beam formula:

$$d_{\text{axial}} \approx w_0 = \frac{\lambda_{\text{i}}}{\pi\ \text{NA}_{\text{exc}}} \tag{2}$$

[1] A popular alternative acronym for LSFM is SPIM, standing for Selective Plane Illumination Microscopy.

with λ_i being the illumination wavelength. It is important to note that the excitation NA is not necessarily the nominal NA of the illumination objective, as in LSFM it is common to underfill the excitation optics, resulting in quite small NAs. The physics of Gaussian beams tells us also the approximate distance where the beam thickness—i.e., the light-sheet thickness—is constant, known as confocal parameter:

$$b = \frac{2\pi\, w_0^2\, n}{\lambda_i} \tag{3}$$

In standard implementations, the confocal parameter is adjusted to be equal to the field of view (FOV) of the microscope. Equation (3) reveals one of the most problematic aspects of LSFM, i.e., the proportionality between light sheet thickness and FOV. When we increase the part of the sample we can image, we simultaneously reduce axial resolution. Several workarounds have been proposed in the last decade, including the use of non-Gaussian beams [26, 27] and axial sweeping of the illumination beam [28, 29].

Uncoupling between illumination and detection has other two important consequences. First, sample-induced aberrations are different along the two optical paths. This implies that the alignment between the illumination light sheet and the detection focal plane can be disrupted if the sample introduces some defocus or deflects illumination light. Automatic focus adjustment is thus important to guarantee high image quality, and several solutions have indeed been proposed in the last years [20, 30]. Second, strong scatterers or absorbers in the sample create the characteristic linear shadowing artifacts, which makes subsequent image analysis challenging [31]. In this case, it should be noted that most of the methods that improve axial resolution also mitigate shadowing, suggesting that standard Gaussian illumination should be overcome in the next future, also in commercial systems.

LSFM: A Game Changer for Neuroscience

During the last two decades, the scientific literature on LSFM has literally exploded (Fig. 1), with numerous advancements on light-sheet generation, image detection, sample mounting, system auto-alignment, and overall design, as reported by many high-level reviews published in recent years [32, 33]. On the other hand, the maturity of this method has been demonstrated by its use in several neuroscientific studies [34]. Most of these studies exploited two fundamental companions of LSFM: tissue processing and automated image analysis.

Imaging of an intact brain requires light to travel into the sample unhindered. Although some vertebrates are naturally transparent in their larval state (e.g., Zebrafish or Medaka fish), biological tissue is usually opaque. Parallel to the development of optical technology, the last 20 years witnessed a great amount of research in clearing protocols [35, 36]. These methods usually include also a whole-mount labeling step, enabling penetration of specific fluorophores inside macroscopic specimens. From an end-user perspective, it is important to underline that clearing and imaging are tightly coupled: the choice of a specific protocol should be evaluated also depending on the available optical instrumentation, both in terms of compatibility of the components and of the signal-to-noise ratio affordable by the microscope [37].

The other technology fundamental in any LSFM laboratory is image processing. As imaging throughput is largely increased, it is easy to generate datasets as large as several TeraBytes. Of course, it is impossible to manually analyze such a large amount of data.

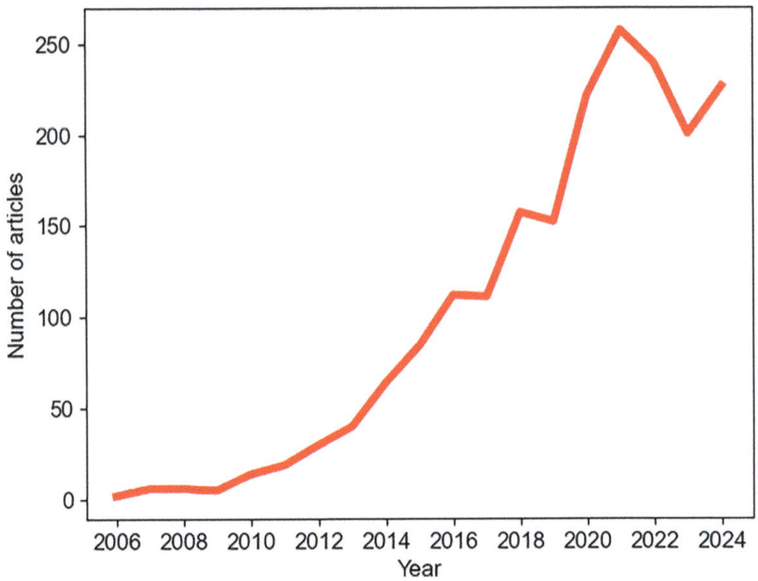

Fig. 1 Growth of scientific papers about LSFM over years. Number of results per year from a query "'light sheet' microscopy" performed on PubMed (https:/pubmed.ncbi.nlm.nih.gov)

Automated solutions for data quantification, maybe exploiting artificial intelligence, are thus becoming popular [38, 39]. But it is also important to establish effective data structures to store images with more than 10^{11} voxels; the effort of the bioimaging community toward development of new standards like OME-Zarr [40] are thus fundamental for a more effective use of the data produced by LSFM.

In conclusion, despite the availability of several commercial solutions, a full adoption of LSFM requires that final users—i.e., neuroscientists—get confident with this methodology and its companions. The community need informed users that are able to distinguish pros and cons of different implementations, and most important, to understand which experimental pipeline is best suited for a specific neuroscientific question. For instance, whole-brain mapping of dense cellular populations requires high-resolution imaging; this in turn entails brilliant staining, some form of autofocus in the microscope and a computational infrastructure capable of digesting several TeraBytes per week. Having all these requirements in mind is a fundamental step for planning experiments or deciding which instrument should be installed.

If we want LSFM to have a significant impact in neuroscience, education of users is a mission as important as technology development. The goal of this book is thus to provide a guide for neuroscientists entering the world of light-sheet fluorescence microscopy. The reader will find a collection of reviews and practical tutorials written by experts in the field, covering both issues in the routine use of the method and recent advancements. The book is organized in three main sections. The first is an introduction to the method, with a chapter by Hans-Ulrich Dodt on LSFM history and a review by Hiroki Ueda et al. on tissue clearing, which is fundamental for imaging macroscopic intact samples. The second part is devoted to some recent advancements in LSFM, aimed at increasing resolution (chapters by Kevin Dean, Ulrich Kubitscheck, Ruixan Gao, and their colleagues), field of view (Adam Glaser), and image contrast (myself). The final section of the book presents several guides to the use of LSFM in an imaging facility (Pages and Batti), for human brain imaging (Costantini et al.), mapping of immediate early genes (Franceschini), and in vivo functional imaging on Zebrafish (Turrini et al.). All the different chapters are targeted at researchers with no expertise in optics, aiming to provide a useful guide to basic and advanced issues.

We are living in the century of the brain. LSFM has the potential to provide us with unique data connecting brain-wide and single-cell approaches, shaping a clearer view of this organ and, ultimately, of ourselves. I hope that this book can be a useful brick in the House of Knowledge that we are building together with past and future generations of scientists.

Sesto Fiorentino (FI), Italy *Ludovico Silvestri*

References

1. Kandel ER, Schwartz JH, Jessell TM (2000) Principles of neural science, 4th edn. McGraw-Hill Companies, Incorporated
2. Yao Z et al (2023) A high-resolution transcriptomic and spatial atlas of cell types in the whole mouse brain. Nature 624(7991):317–332. https://doi.org/10.1038/s41586-023-06812-z
3. Komendantov AO, Venkadesh S, Rees CL, Wheeler DW, Hamilton DJ, Ascoli GA (2019) Quantitative firing pattern phenotyping of hippocampal neuron types. Sci Rep 9(1):17915. https://doi.org/10.1038/s41598-019-52611-w
4. Kanari L et al (2019) Objective morphological classification of neocortical pyramidal cells. Cereb Cortex 29(4):1719–1735. https://doi.org/10.1093/cercor/bhy339
5. Ascoli GA et al (2008) Petilla terminology: nomenclature of features of GABAergic interneurons of the cerebral cortex. Nat Rev. Neurosci 9(7):557–568. https://doi.org/10.1038/nrn2402.
6. Gouwens NW et al (2019) Classification of electrophysiological and morphological neuron types in the mouse visual cortex. Nat Neurosci 22(7):1182–1195. https://doi.org/10.1038/s41593-019-0417-0
7. Peng H et al (2021) Morphological diversity of single neurons in molecularly defined cell types. Nature 598(7879):174–181. https://doi.org/10.1038/s41586-021-03941-1
8. Ramón y Cajal S (1899) Textura del sistema nervioso del hombre y de los vertebrados: estudios sobre el plan estructural y composición histológica de los centros nerviosos adicionados de consideraciones fisiológicas fundadas en los nuevos descubrimientos, vol I. Nicolás Moya, Madrid. Consultato: 10 maggio 2025. [Online]. Disponibile su: https://digibug.ugr.es/handle/10481/69713
9. Hodgkin AL, Katz B (1949) The effect of sodium ions on the electrical activity of the giant axon of the squid. J Physiol 108(1):37–77. https://doi.org/10.1113/jphysiol.1949.sp004310
10. Hodgkin AL, Huxley AF (1952) A quantitative description of membrane current and its application to conduction and excitation in nerve. J Physiol 117(4):500–544. https://doi.org/10.1113/jphysiol.1952.sp004764
11. Tasic B et al (2018) Shared and distinct transcriptomic cell types across neocortical areas. Nature 563(7729):72–78. https://doi.org/10.1038/s41586-018-0654-5
12. Brodmann K (1985) Vergleichende Lokalisationslehre der Grosshirnrinde: in ihren Prinzipien dargestellt auf Grund d. Zellenbaues, Repr. d. 1. Aufl. von 1909. Barth, Leipzig
13. Dolan RJ (2008) Neuroimaging of cognition: past, present, and future. Neuron 60(3):496–502. https://doi.org/10.1016/j.neuron.2008.10.038
14. Poldrack RA (2008) The role of fMRI in Cognitive Neuroscience: where do we stand? Curr Opin Neurobiol 18(2):223–227. https://doi.org/10.1016/j.conb.2008.07.006

15. Huettel SA (2012) Event-related fMRI in cognition. NeuroImage 62(2):1152–1156. https://doi.org/10.1016/j.neuroimage.2011.08.113
16. Le Bihan D (2003) Looking into the functional architecture of the brain with diffusion MRI. Nat Rev. Neurosci 4(6):469–480. https://doi.org/10.1038/nrn1119
17. Bassett DS, Sporns O (2017) Network neuroscience. Nat Neurosci 20(3):353–364. https://doi.org/10.1038/nn.4502
18. Yuste R (2015) From the neuron doctrine to neural networks. Nat Rev. Neurosci 16(8):487–497. https://doi.org/10.1038/nrn3962
19. Ueda HR, Dodt H-U, Osten P, Economo MN, Chandrashekar J, Keller PJ (2020) Whole-brain profiling of cells and circuits in mammals by tissue clearing and light-sheet microscopy. Neuron 106(3):369–387. https://doi.org/10.1016/j.neuron.2020.03.004
20. Silvestri L et al (2021) Universal autofocus for quantitative volumetric microscopy of whole mouse brains. Nat Methods 18(8):953–958. https://doi.org/10.1038/s41592-021-01208-1
21. Ahrens MB, Orger MB, Robson DN, Li JM, Keller PJ (2013) Whole-brain functional imaging at cellular resolution using light-sheet microscopy. Nat Methods 10(5):413–420. https://doi.org/10.1038/nmeth.2434
22. Pawley JB (2006) Handbook of biological confocal microscopy. Springer US, Boston, MA. https://doi.org/10.1007/978-0-387-45524-2
23. Zipfel WR, Williams RM, Webb WW (2003) Nonlinear magic: multiphoton microscopy in the biosciences. Nat Biotechnol 21(11):1369–1377. https://doi.org/10.1038/nbt899
24. Silvestri L, Bria A, Sacconi L, Iannello G, Pavone FS (2012) Confocal light sheet microscopy: micron-scale neuroanatomy of the entire mouse brain. Opt Express 20(18):20582–20598. https://doi.org/10.1364/OE.20.020582
25. Müllenbroich MC et al (2015) Comprehensive optical and data management infrastructure for high-throughput light-sheet microscopy of whole mouse brains. Neurophotonics 2(4):041404. https://doi.org/10.1117/1.NPh.2.4.041404
26. Fahrbach FO, Simon P, Rohrbach A (2010) Microscopy with self-reconstructing beams. Nat Photonics 4(11):780–785. https://doi.org/10.1038/nphoton.2010.204
27. Vettenburg T et al (2014) Light-sheet microscopy using an Airy beam. Nat Methods 11(5):541–544. https://doi.org/10.1038/nmeth.2922
28. Chakraborty T et al (2019) Light-sheet microscopy of cleared tissues with isotropic, subcellular resolution. Nat Methods 16(11):1109–1113. https://doi.org/10.1038/s41592-019-0615-4
29. Voigt FF et al (2019) The mesoSPIM initiative: open-source light-sheet microscopes for imaging cleared tissue. Nat Methods 16(11):1105–1108. https://doi.org/10.1038/s41592-019-0554-0
30. Tomer R, Ye L, Hsueh B, Deisseroth K (2014) Advanced CLARITY for rapid and high-resolution imaging of intact tissues. Nat Protoc 9(7):1682–1697. https://doi.org/10.1038/nprot.2014.123
31. Ricci P et al (2022) Removing striping artifacts in light-sheet fluorescence microscopy: a review. Prog Biophys Mol Biol 168:52–65. https://doi.org/10.1016/j.pbiomolbio.2021.07.003
32. Stelzer EHK et al (2021) Light sheet fluorescence microscopy. Nat Rev. Methods Primer 1(1):1–25. https://doi.org/10.1038/s43586-021-00069-4
33. Daetwyler S, Fiolka RP (2023) Light-sheets and smart microscopy, an exciting future is dawning. Commun Biol 6(1):1–11. https://doi.org/10.1038/s42003-023-04857-4
34. Hillman EMC, Voleti V, Li W, Yu H (2019) Light-sheet microscopy in neuroscience. Annu Rev. Neurosci 42:295–313. https://doi.org/10.1146/annurev-neuro-070918-050357
35. Richardson DS, Lichtman JW (2015) Clarifying tissue clearing. Cell 162(2):246–257. https://doi.org/10.1016/j.cell.2015.06.067
36. Ueda HR et al. (2020) Tissue clearing and its applications in neuroscience. Nat Rev. Neurosci 21(2):61–79. https://doi.org/10.1038/s41583-019-0250-1
37. Silvestri L, Costantini I, Sacconi L, Pavone FS (2016) Clearing of fixed tissue: a review from a microscopist's perspective. J Biomed Opt 21(8):081205. https://doi.org/10.1117/1.JBO.21.8.081205

38. Berg S et al (2019) ilastik: interactive machine learning for (bio)image analysis. Nat Methods 16(12): 12. https://doi.org/10.1038/s41592-019-0582-9
39. Ruan X et al (2024) Image processing tools for petabyte-scale light sheet microscopy data. Nat Methods 21(12):2342–2352. https://doi.org/10.1038/s41592-024-02475-4
40. Moore J et al (2021) OME-NGFF: a next-generation file format for expanding bioimaging data-access strategies. Nat Methods 18(12):1496–1498. https://doi.org/10.1038/s41592-021-01326-w

Contents

Contributors

FUMITO AKIYAMA • *Laboratory for Synthetic Biology, Center for Biosystems Dynamics Research, RIKEN, Osaka, Japan; Department of Ophthalmology and Visual Sciences, Graduate School of Biomedical Sciences, Nagasaki University, Nagasaki, Japan*

LAURA BATTI • *Wyss Center for Bio and Neuro Engineering, Geneva, Switzerland*

HAZEL M. BORGES • *Lyda Hill Department of Bioinformatics, UT Southwestern Medical Center, Dallas, TX, USA; Cecil H. and Ida Green Center for Systems Biology, UT Southwestern Medical Center, Dallas, TX, USA*

SAMANTHA CALVACHE • *Lyda Hill Department of Bioinformatics, UT Southwestern Medical Center, Dallas, TX, USA; Cecil H. and Ida Green Center for Systems Biology, UT Southwestern Medical Center, Dallas, TX, USA*

IRENE COSTANTINI • *Department of Biology, University of Florence, Florence, Italy; European Laboratory for Non-linear Spectroscopy (LENS), Sesto Fiorentino, Italy*

GIUSEPPE DE VITO • *European Laboratory for Non-linear Spectroscopy (LENS), Sesto Fiorentino, Italy; Department of Neuroscience, Psychology, Drug Research and Child Health, University of Florence, Florence, Italy*

KEVIN M. DEAN • *Lyda Hill Department of Bioinformatics, UT Southwestern Medical Center, Dallas, TX, USA; Cecil H. and Ida Green Center for Systems Biology, UT Southwestern Medical Center, Dallas, TX, USA*

HANS-URICH DODT • *University of Vienna, Faculty of Life Sciences, Department of Pharmaceutical Chemistry, MMCT-lab, Josef-Holaubeck-Platz, Vienna, Austria; Medical University Vienna, Center for Brain Research, Section of Bioelectronics, Vienna, Austria*

ALESSANDRA FRANCESCHINI • *Department of Psychiatry, Columbia University Irving Medical Center (CUIMC), New York, NY, USA; European Laboratory for Non-linear Spectroscopy (LENS), Sesto Fiorentino, Italy*

RUIXUAN GAO • *Department of Chemistry, University of Illinois Chicago, Chicago, IL, USA; Department of Biological Sciences, University of Illinois Chicago, Chicago, IL, USA*

ADAM GLASER • *Allen Institute for Neural Dynamics, Seattle, WA, USA*

JOHN HAUG • *Lyda Hill Department of Bioinformatics, UT Southwestern Medical Center, Dallas, TX, USA; Cecil H. and Ida Green Center for Systems Biology, UT Southwestern Medical Center, Dallas, TX, USA*

ULRICH KUBITSCHECK • *Clausius Institute of Physical and Theoretical Chemistry, University of Bonn, Bonn, Germany*

HELEN C. LAI • *Department of Neuroscience, UT Southwestern Medical Center, Dallas, TX, USA*

SHUANG LIANG • *Department of Immunology, UT Southwestern Medical Center, Dallas, TX, USA*

JINLONG LIN • *Lyda Hill Department of Bioinformatics, UT Southwestern Medical Center, Dallas, TX, USA; Cecil H. and Ida Green Center for Systems Biology, UT Southwestern Medical Center, Dallas, TX, USA*

XUEMEI LUO • *Department of Immunology, UT Southwestern Medical Center, Dallas, TX, USA*

KATSUHIKO MATSUMOTO • *Laboratory for Synthetic Biology, Center for Biosystems Dynamics Research, RIKEN, Osaka, Japan; Department of Systems Pharmacology, University of Tokyo, Tokyo, Japan*
STÉPHANE PAGÈS • *Wyss Center for Bio and Neuro Engineering, Geneva, Switzerland*
FRANCESCO SAVERIO PAVONE • *National Institute of Optics, National Research Council, Sesto Fiorentino, Italy; European Laboratory for Non-linear Spectroscopy (LENS), Sesto Fiorentino, Italy; Department of Physics and Astronomy, University of Florence, Sesto Fiorentino, Italy*
LUCA PESCE • *Center for Instrument Sharing of the University of Pisa (CISUP), University of Pisa, Pisa, Italy*
ILIODORA V. POP • *Department of Neuroscience, UT Southwestern Medical Center, Dallas, TX, USA*
JUAN EDUARDO RODRIGUEZ-GATICA • *Clausius Institute of Physical and Theoretical Chemistry, University of Bonn, Bonn, Germany; Institute for Experimental Epileptology and Cognition Research (EECR), University of Bonn Medical School, Bonn, Germany*
MARTIN KARL SCHWARZ • *Institute for Experimental Epileptology and Cognition Research (EECR), University of Bonn Medical School, Bonn, Germany*
LUDOVICO SILVESTRI • *Department of Physics and Astronomy, University of Florence, Sesto Fiorentino, FI, Italy; European Laboratory for Non-linear Spectroscopy (LENS), Sesto Fiorentino, FI, Italy*
LAPO TURRINI • *National Institute of Optics, National Research Council, Sesto Fiorentino, Italy; European Laboratory for Non-linear Spectroscopy (LENS), Sesto Fiorentino, Italy*
HIROKI R. UEDA • *Laboratory for Synthetic Biology, Center for Biosystems Dynamics Research, RIKEN, Osaka, Japan; Department of Systems Pharmacology, University of Tokyo, Tokyo, Japan; Department of Systems Biology, Kurume University, Kurume, Japan*
WEI WANG • *Department of Chemistry, University of Illinois Chicago, Chicago, IL, USA*
ZHENYU ZHONG • *Department of Immunology, UT Southwestern Medical Center, Dallas, TX, USA*
FELIX Y. ZHOU • *Lyda Hill Department of Bioinformatics, UT Southwestern Medical Center, Dallas, TX, USA; Cecil H. and Ida Green Center for Systems Biology, UT Southwestern Medical Center, Dallas, TX, USA*

Part I

Light-Sheet Fluorescence Microscopy (LSFM): A General Introduction

Chapter 1

The Development of Ultramicroscopy

Hans-Urich Dodt

Abstract

Light-sheet fluorescence microscopy has become a very popular method for imaging biological specimens. The very first implementation of a microscope based on planar illumination was the ultramicroscope, invented at the beginning of the twentieth century. Starting from this first instrument, this chapter reviews the history and development of light-sheet microscopy, up to the most recent advancements.

Key words Ultramicroscopy, Light-sheet microscopy history, Glass brain

1 Introduction

The first use of ultramicroscopy, here used as synonym for light-sheet microscopy of cleared samples, was reported by Siedentopf and Zsigmondy [1] (Fig. 1). Siedentopf was an optical physicist working on problems in chemistry, specifically the visualization of small colloidal particles (ultramicronen). At that time, no powerful darkfield condensers were available, so Siedentopf and colleagues came up with the idea to illuminate their colloidal solutions from the side to detect only the scattered light. In order to illuminate a well-defined volume, they created a sheet of light by using lenses, aperture stops, and microscope objectives. In the year 1900, lasers were not yet available, and therefore, a bulky carbon arc lamp was used. Although the ultramicroscope was marketed by Zeiss, it was never used widely and its application remained confined to colloidal chemistry. Further efforts in microscope development led instead to the construction of efficient darkfield condensers such as the cardioid condenser, which allowed the generation of darkfield illumination with numerical apertures of up to 1.4. Transmitted darkfield microscopy with high numerical aperture condensers employs inclined ring-like light-sheets with an oblique plane to the horizontal. The obliqueness is not a problem as long as the aim of the illumination is exclusively the recording of scattered light and not optical sectioning. Before the advent of confocal microscopy,

Ludovico Silvestri (ed.), *Light Sheet Microscopy*, Neuromethods, vol. 229, https://doi.org/10.1007/978-1-0716-5100-1_1,

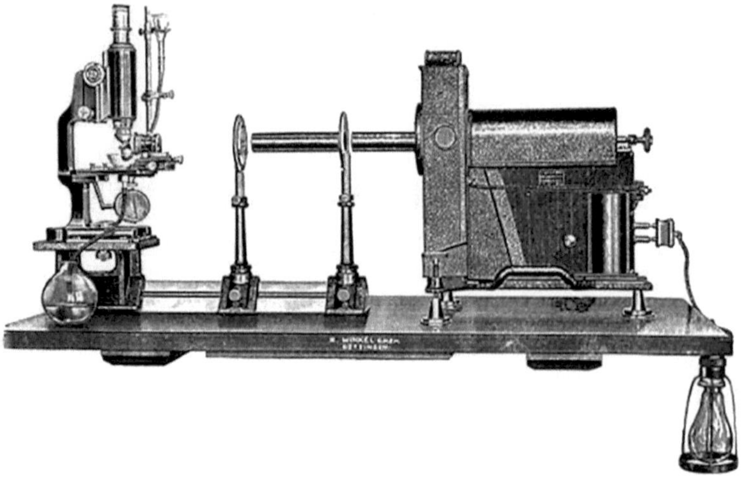

Fig. 1 First ultramicroscope developed by Siedentopf and Zsigmondy. (Adapted from Siedentopf and Zsigmondy [1])

darkfield microscopy with incident light was used in the 1960s in biological research and marketed as "Ultropak" by Leitz. As it allowed at least some kind of optical sectioning in contrast to full-field incident light microscopy, it was used to visualize blood vessels and erythrocyte streaming in the kidney of rats, for example [2].

2 First Ultramicroscope Developed by Siedentopf and Zsigmondy [1]

In other fields, real light-sheet microscopy was used, but only for surface imaging, most prominently in material macrography. The light-sheet was employed to visualize the surface of minerals and insect eyes [3]. A similar approach was used in light scanning photomicrography for the visualization of the surface of insects with two-sided or three-sided illumination [4]. The instrument was primarily designed as an imaging tool with lower magnification, complementary to the scanning electron microscope. A similar arrangement was described for the macroscopic imaging of screws, coins, and rifle bullets [5]. The use of laser light-sheets was then reported in the context of fluid dynamics investigations [6]. However, nearly all these approaches relied on light-sheet microscopy mainly as a means to visualize the surface of objects. This way, light-sheet microscopy had no important impact in biology as the surface of biological objects is often not of particular interest.

There was a simple reason for only imaging surfaces: Biological specimens with more than a few cell layers are mostly opaque, and when light manages to proceed below the surface, stray light is immediately generated which quickly obscures the images of deeper layers in the object. The attempt to exclude stray light in the confocal microscope quickly comes to its limits if specimens are thicker than a few tens of micrometers due to strong scattering by dense cell layers. To be generally applicable, a new idea was needed for light-sheet microscopy and came from a totally different field of science: chemistry. The basic idea was that if a biological object was not transparent, it had to be made transparent. In principle, it was already possible to make biological tissue transparent more than 100 years ago.

The first reports on clearing opaque biological samples date back about 120 years, similar to the first reports on light-sheet microscopy. At that time, anatomists tried to render fixed anatomical preparations translucent. It was mainly the German anatomist Walter Spalteholtz who pursued these studies in a systematic fashion [7]. He worked in Leipzig at the anatomy institute of Wilhelm His, who was interested in the nervous system of the heart. Spalteholtz studied the blood vessels of the heart, which were of particular interest with respect to infarcts during heart attacks. He tested numerous organic solutions in order to obtain optimally transparent preparations. A mixture of benzyl alcohol and methyl salicylate was found by him to be most effective for clearing large specimens. Spalteholtz had a number of competitors as some colleagues tried similar approaches. Large preparations in the decimeter range [8] were eventually cleared by them with methods generating a terrible smell.

Spalteholtz's method made no strong scientific impact at that time, as simply looking at cleared specimens does not really allow quantification of the object. In principle, quantitative imaging would have been possible with a macroscopic light-sheet microscope using a carbon arc as light source applied to specimens like the macroscopic human fetuses that Spalteholz cleared with his one-year-long chemical procedure. This would certainly have been of scientific interest as human embryogenesis was studied at that time by preparing serial macroscopic sections. For a 3D impression of the serial sections, wax plates with the circumferences of the sections were stacked on top of each other [9]. With a "Makro-Ultramikroskop," this task could have been achieved much more easily without sectioning. Photos of the optical sections in the ultramicroscope could have been taken, printed on transparencies, and stacked above each other for a 3D reconstruction. But obviously, the time had not yet come for this idea.

Actually, it was embryologists who first used clearing to address biological questions [10]. For embryo imaging, a special clearing solution was developed, a mixture of benzyl alcohol and benzyl benzoate (Murray's clear), replacing the methyl salicylate in Spalteholtz's recipe. This technique was then used infrequently by surgeons and anatomists for angiographic investigations in large preparations like the human larynx.

The first one to use clearing of biological specimens in conjunction with a quantitative method was James Sharpe. He combined clearing with optical projection tomography [11]. Whole mouse embryos were imaged using an optical method mathematically similar to X-ray computer tomography but using visible light. 3D visualization was realized by trans-illumination imaging of the embryos from multiple directions, achieving a resolution of around 20 μm. Afterwards, algorithms known from CT X-ray imaging were used to generate the 3D reconstruction. The 3D images were obtained by staining the embryos with absorbing or fluorescent dyes; imaging of endogenous proteins like GFP was not yet described.

Imaging of GFP or other endogenous fluorescent protein-expressing tissue was not possible at this time, as GFP was destroyed by chemical clearing processes. For this reason, the first attempts to image entire whole-mount preparations without sectioning suffered from low spatial resolution and challenges in using fluorescent proteins.

Light-sheet microscopy was used in biology for the first time by Arno Voie in the lab of Spelman [12]. They used light-sheet illumination for fluorescence excitation in the guinea pig cochlea previously immersed in fluorescent rhodamine. The cochlea was made transparent before by Spalteholtz's clearing. With this approach, it was possible to obtain reconstructions of the anatomy of the inner ear and extract quantitative data, which would have been difficult with standard histological methods. Although all basic elements of modern light-sheet microscopy were there, the method remained largely unnoticed. The reason was probably that the authors had no access to modern 3D reconstruction software, although such programs were available at that time for confocal microscopy. The custom-made software of the Spelman lab yielded rather unattractive wire grid visualizations. An application of optical clearing to more interesting biological specimens was not realized by them. So, another decade passed before light-sheet microscopy in combination with clearing was used again in biology.

Before clearing came into play, light-sheet microscopy, also known as selective plane illumination microscopy (SPIM), was applied to some of the rare semitransparent biological specimens by Ernst Stelzer's lab [13]. By imaging their specimens from different sides with light-sheet microscopy and then combining the recordings, they could even image zebrafish development over time.

3 First Cleared Mouse Brain Presented [14]

The solution for the problem of 3D imaging of opaque biological specimens like the mouse brain was provided by Hans-Ulrich

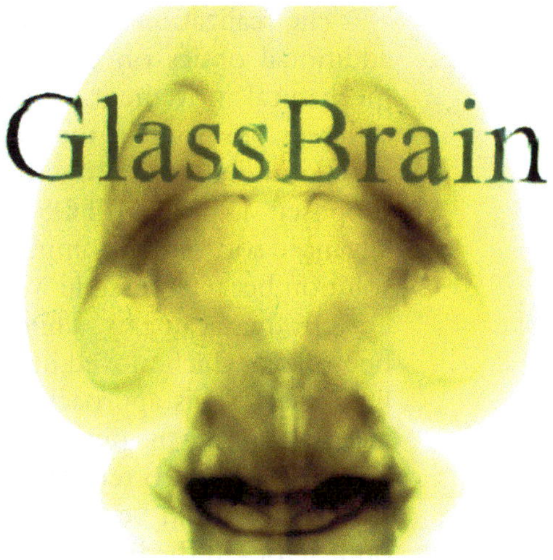

Fig. 2 First cleared mouse brain presented. (Adapted from Dodt et al. [14])

Dodt's lab [14] (Fig. 2). By applying light-sheet microscopy to brains expressing GFP, a dense network of neurons in the hippocampus became visible [15]. The mouse strains used expressed GFP in subsets of neurons controlled by the *Thy1* promotor. By modification of Murray's clearing and imaging immediately after the mouse brains became transparent, it was possible to image neuronal networks in entire mouse brains. Instead of the multiple view imaging for semitransparent living specimens for these cleared objects, it was sufficient to record only one image stack with simultaneous two-sided illumination. Using this approach in excised hippocampi, dendritic trees of neurons with small structures resembling dendritic spines could be recorded. The interior of entire adult *Drosophila melanogaster* was visualized by autofluorescence after the flies were cleared with a similar method as the mouse brains [16]. Some years later, the stability of the GFP in cleared brains could be significantly improved by adjustments of the clearing chemistry [17] which started the development of a multitude of similar organic optical clearings becoming popular as "DISCO" clearings.

Regarding the microscopy technique, Francesco Pavone's group realized further enhancements in image contrast by applying the principle of confocality to light-sheet microscopy [18]. In their study, line confocality was realized by a mechanical slit on the detection side on to which the sidewise-moving thinnest part of the light-sheet was reimaged. In another publication from Stelzer's lab, the light-sheet was generated by a fast-moving laser beam [19]. A further improvement of light-sheet microscopy came from Jan Huisken by wobbling the laser beam at a small angle in the light-sheet plane to avoid the generation of stripe artifacts occurring in typical light-sheet microscopy recordings [20].

The realization of confocality with a mechanical slit required additional optics on the detection side. An alternative was the mechanical movement of the thinnest part of the light-sheet to different positions in the field of view and stitching the sharpest parts of these images together to a final crisp image. This was first proposed as HR-OPFOS [21] and realized later with more advanced software in commercial ultramicroscopes. A more elegant solution became possible with the advent of modern CMOS cameras. In these cameras, a line of active pixels simulating a slit could be realized which moved in synchrony with the thinnest part of the light-sheet. This approach obviated the need to stitch the image together as only the pixels opposite to the sharpest part of the light-sheet were used to record the image. The movement of the light-sheet focus could be realized elegantly by moving it with electro-optic lenses in the light-sheet generator [22].

However, both approaches, stitching and slit scanning, result in longer exposure times of the specimen than a single exposure to a light-sheet with a longer thin part (increased Rayleigh range) would need. This is critical as many fluorophores, especially widely used endogenous fluorophores like GFP, bleach easily. Several attempts were made to realize such a long light-sheet with nondiffracting beams like Bessel [23] or Airy [24] beams. Both Bessel and Airy beams suffer from considerably greater sidelobes and are thus not widely used besides special applications. A partial escape from this dilemma can be realized by the interference of many Bessel beams side by side (lattice light-sheet) for small preparations [25]. Another promising attempt is the combination of Gaussian and mesooptic elements leading to an increased Rayleigh range at a very moderate cost of sidelobes [26]. A completely different approach for suppression of sidelobes is nonlinear optics using the two-photon effect used in 2P scanning microscopes [27]. However, this comes at the cost of an increased energy exposure and difficulties to change the excitation wavelength to excite specifically different fluorophores.

So, one can say that the ideal light-sheet microscope has yet not been realized. However, in the meantime, ultramicroscopy as light-sheet microscopy of cleared specimens is widely used for imaging of various biological and medical objects. As light-sheet microscopy is much easier to realize than confocal microscopy, one is wondering why it was not yet developed decades ago. It seems that sometimes a simple idea is sufficient to open a new field in science.

Besides higher resolution, the development is now going to ever larger specimens. For this special application, "mammoth" light-sheet microscopes are in development using super large lenses normally used for chip inspection in solid state physics [28]. For light-sheet microscopy of human brain tissue only, the use of millimeter-thick slices has been reported up to now [29]. However, this approach struggles with alignment problems of the recordings and uneven shrinkage of the sections. However, if one looks at the explosive development of ultramicroscopy in the last 15 years, one is tempted to predict a whole cleared human brain is not out of reach 1 day.

Conflict of Interest

The authors have no conflicts of interest to declare that are relevant to the content of this chapter.

References

1. Siedentopf H, Zsigmondy R (1903) Über Sichtbarmachung und Größenbestimmung ultramikroskopischer Teilchen, mit besonderer Anwendung auf Goldrubingläser. Ann Phys 315:1–39
2. Steinhausen M, Tillmanns H, Thederan H (1978) Microcirculation of the epimyocardial layer of the heart. I. A method for in vivo observation of the microcirculation of superficial ventricular myocardium of the heart and capillary flow pattern under normal and hypoxic conditions. Pflugers Arch 378:9–14
3. Mclachlan D (1964) Extreme focal depth in microscopy. Appl Opt 3:1009
4. Schram R (1981) Light-scanning photomicrography. Microscope 29:13–17
5. Root N (1991) A simplified unit for making deep-field (scanning) macrographs. J Biol Photography 59:3–8
6. Decre M, Buchlin JM (1994) An extra-thin light-sheet technique used to investigate meniscus shapes by laser-induced fluorescence. Exp Fluids 16:339–341
7. Spalteholz W (1914) Über das Durchsichtigmachen von menschlichen und tierischen Präparaten, S. Hierzel, Leipzig
8. Lundvall H (1905) Weiteres über Demonstration embryonaler Skelette. Anat Anz 27:520–523
9. Nauck ET (1942) Über "Lageentwicklung" und "Entwicklung von Körpergegenden". Ber Naturforsch Ges Freib i Br 38:109–116
10. Dent JA, Polson A, Klymkowsky MW (1989) A whole-mount immunocytochemical analysis of the expression of the intermediate filament protein vimentin in Xenopus. Development 105:61–74
11. Sharpe J, Ahlgren U, . Perry, Hill B, Ross A, Hecksher-Sorensen, Baldock R, Davidson (2002) Optical projection tomography as a tool for 3D microscopy and gene expression studies. Science, 296, pp. 541–545
12. Voie AH, Burns DH, Spelman FA (1993) Orthogonal-plane fluorescence optical sectioning: three-dimensional imaging of macroscopic biological specimens. J Microsc 170:229–236
13. Huisken J, Swoger J, Del Bene F, Wittbrodt J, Stelzer EH (2004) Optical sectioning deep inside live embryos by selective plane illumination microscopy. Science 305:1007–1009
14. Dodt HU, Schierloh A, Leischner U, Becker K, Eder M, Zieglgänsberger W (2005) The transparent brain: visualization of neuronal networks in the whole mouse brain. Meeting Society for Neuroscience 2005, Abstract 705.3
15. Dodt HU, Leischner U, Schierloh A, Jahrling N, Mauch CP, Deininger, Deussing JM, Eder M, Zieglgänsberger W, Becker K (2007) Ultramicroscopy: three-dimensional visualization of neuronal networks in the whole mouse brain. Nat Methods 4:331–336
16. Jahrling N, Becker K, Schonbauer C, Schnorrer F, Dodt HU (2010) Three-dimensional reconstruction and segmentation of intact Drosophila by ultramicroscopy. Front Neurosci 4:1–23
17. Becker K, Jährling N, Saghafi S, Weiler R, Dodt HU (2012) Chemical clearing and dehydration of GFP expressing mouse brains. PLoS One 7: e33916
18. Silvestri L, Bria A, Sacconi L, Iannello G, Pavone FS (2012) Confocal light-sheet microscopy: micron-scale neuroanatomy of the entire mouse brain. Opt Express 20(18): 20582–20598
19. Keller PJ, Stelzer EH (2010) Digital scanned laser light-sheet fluorescence microscopy. Cold Spring Harb Protoc 2010(5):pdb.top78
20. Huisken J, Stainier DY (2007) Even fluorescence excitation by multidirectional selective

plane illumination microscopy (mSPIM). Opt Lett 32:2608–2610
21. Buytaert A, Descamps E, Adriaens D, Dirckx JJ (2012) The OPFOS microscopy family: high-resolution optical sectioning of biomedical specimens. Anat Res Int 2012:206238
22. Fiolka R, Stemmer A, Belyaev Y (2007) Virtual slit scanning microscopy. Histochem Cell Biol 128:499–505
23. Fahrbach FO, Rohrbach A (2012) Propagation stability of self-reconstructing Bessel beams enables contrast-enhanced imaging in thick media. Nat Commun 3:632
24. Vettenburg T, Dalgarno H, Nylk J, Coll-Lladó C, Ferrie DEK, Čižmár T, Gunn-Moore FJ, Dholakia K (2014) Light-sheet microscopy using an Airy beam. Nat Methods 11:541–544
25. Chen BC, Legant WR, Wang K, Shao L, Milkie DE, Davidson MW, Janetopoulos C, Wu XS, Hammer JA, Liu Z, English BP, Mimori-Kiyosue Y, Romero DP, Ritter AT, Lippincott-Schwartz J, Fritz-Laylin, Mullins R, Mitchell DM, Bembenek JN, Reymann AC, Böhme R, Grill SW, Wang JT, Seydoux G, Tulu US, Kiehart DP, Betzig E (2014) Lattice light-sheet microscopy: imaging molecules to embryos at high spatiotemporal resolution. Science 346(6208):1257998
26. Saghafi S, Becker K, Hahn C, Dodt HU (2014) 3D-ultramicroscopy utilizing aspheric optics. J Biophotonics 7:117–125
27. Truong TV, Supatto W, Koos DS, Choi JM, Fraser SE (2011) Deep and fast live imaging with two-photon scanned light-sheet microscopy. Nat Methods 8:757–760
28. Glaser A, Chandrashekar J, Vasquez J, Arshadi C, Ouellette N, Jiang X, Baka J, Kovacs G, Woodard M, Seshamani S, Cao K, Clack N, Recknagel A, Grim A, Balaram P, Turschak E, Liddell A, Rohde J, Hellevik A, Takasaki K, Barner LE, Logsdon M, Chronopoulos C, de Vries S, Ting J, Perlmutter S, Kalmbach B, Dembrow N, Reid RC, Feng D, Svoboda K (2023) Expansion-assisted selective plane illumination microscopy for nanoscale imaging of centimeter-scale tissues. Elife. https://doi.org/10.7554/eLife.91979.1
29. Costantini I, Morgan L, Yang J, Balbastre Y, Varadarajan D, Pesce L, Scardigli M, Mazzamuto G, Gavryusev V, Castelli M, Roffilli M, Silvestri L, Laffey J, Raia S, Varghese M, Wicinski B, Chang S, Chen IA, Wang H, Cordero D, Vera M, Nolan J, Nestor K, Mora J, Iglesias JE, Garcia Pallares E, Evancic K, Augustinack JC, Fogarty M, Dalca AV, Frosch MP, Magnain C, Frost R, van der Kouwe A, Chen SC, Boas DA, Pavone FS, Fischl B, Hof PR (2023) A cellular resolution atlas of Broca's area. Sci Adv 9:eadg3844

Chapter 2

Tissue Clearing

Fumito Akiyama, Katsuhiko Matsumoto, and Hiroki R. Ueda

Abstract

Tissue-clearing technologies are a fundamental companion to light-sheet fluorescence microscopy. These methods enable whole-organ transparency by removing lipids and pigments and homogenizing refractive index. In this chapter, we review the main principles underlying tissue clearing and describe an end-to-end pipeline for analyzing entire mouse brains with the CUBIC-Atlas and the CUBIC-Cloud platforms.

Key words Tissue clearing, Whole-mount staining, Whole-brain imaging, CUBIC

1 Tissue-Clearing Technology and Principles

Although tissue clearing was developed more than 100 years ago, this method has only recently been combined with light-sheet microscopy. Lundvall in Sweden and Spalteholz in Germany first made tissue transparent using hydrophobic reagents based on methyl salicylate and benzyl benzoate [1, 2]. Tuchin of Russia developed tissue-clearing reagents comprised of hydrophobic chemicals in 1995 (including X-ray contrast, alcohols such as glycerol, sugars such as glucose, and DMSO) [3–5]. In 2013, Deisseroth, Chung, and colleagues published a hydrogel embedding tissue-clearing technique called "Clarity" [6]. Each of these tissue-clearing techniques has advantages and disadvantages, which should be understood to select the most appropriate technique for each study. For example, hydrophobic clearing offers rapid and scalable transparency, but low retention of fluorescent proteins and organic solvent hazards [7–10]. Hydrophilic reagents retain fluorescent proteins and are safe, but they require more time and achieve less transparency than organic solvents [11–14]. Hydrogel embedding-based clearing also retains fluorescent proteins and achieves high transparency, but it requires special equipment and has limited scalability.

Ludovico Silvestri (ed.), *Light Sheet Microscopy*, Neuromethods, vol. 229, https://doi.org/10.1007/978-1-0716-5100-1_2,

Improvements in tissue-clearing techniques have been developed recently, including delipidation with stronger delipidation agents, enhanced delipidation conditions (electrophoresis and ultrasound), and improved refractive index (RI) adjusting reagents [6, 15]. Transparency has been particularly difficult to achieve in the human tissues because strong tissue clearing is necessary, but Murray and colleagues reported the application of SWITCT in 2015 and the application of CUBIC (Clear, Unobstructed Brain/Body Imaging Cocktails and Computational Analysis) and 1,2-hexanediol as a delipidation reagent in 2019 to achieve transparency in human brain blocks [16–18]. In 2020, a whole human brain clearing technique (SHANEL) was reported, but many issues still need to be addressed, including increasing transparency and practical throughput [19].

The combination of tissue-clearing techniques, 3D immunostaining, and 3D imaging is useful for advancing biological studies and disease elucidations but is still in the development phase, and further development is expected in each of these techniques. Herein, we review tissue clearing and 3D staining techniques and their applications (Fig. 1), with a focus on CUBIC as an example.

1.1 Delipidation

Biological specimens have multiple components, including water, proteins, and lipids, with different RIs. RI heterogeneity causes

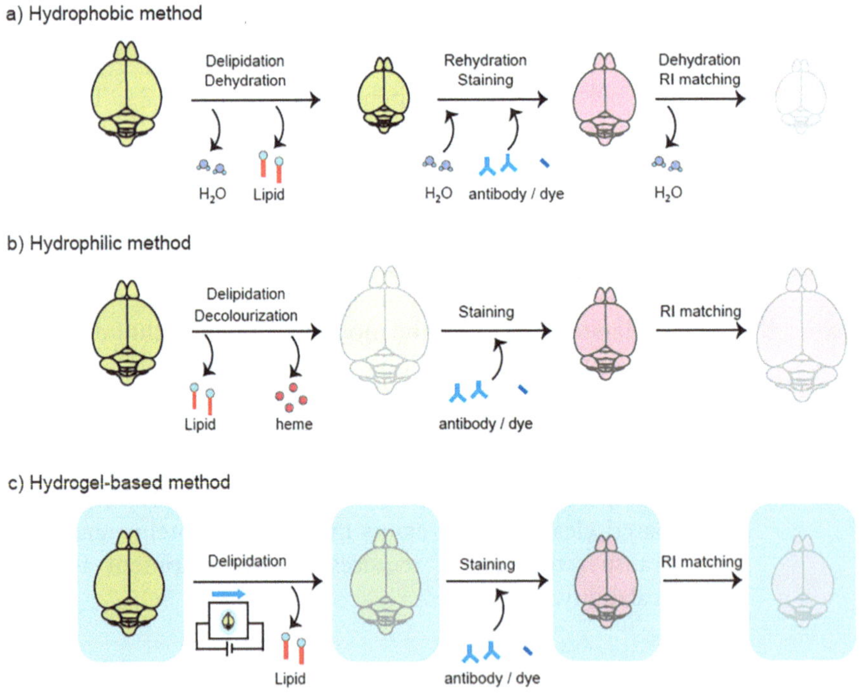

Fig. 1 Overview of typical tissue-clearing and staining methods. (**a**) Hydrophobic solvent-based tissue clearing method. (**b**) Hydrophilic solution-based tissue clearing method. (**c**) Hydrogel-based tissue clearing method

light scattering, leading to optical opacity. Delipidation, one of the steps in tissue clearing, reduces RI heterogeneity in biological specimens by removing lipids and improves the diffusion of reagents into the specimen during staining by creating holes in the tissue and cell membranes. Of note, cell-cell adhesion cannot be maintained after delipidation. Thus, fixation by cross-linking reagents such as PFA or formalin is essential before delipidation. Traditionally, delipidation with solvents, such as methanol, ethanol, or THF, is common, but these organic solvents can quench fluorescent proteins and deform tissue due to dehydration. CUBIC-L, which is composed of Triton X-100 and *N*-butyl-diethanolamine, was developed by screening more than 1600 hydrophilic compounds [20]. Achieving high transparency in primate brains has been challenging, but hexanediol was discovered by screening compounds for delipidation of human brain white matter [17]. Although delipidation with hydrophilic compounds is time-consuming, this method is superior to delipidation with organic solvents for preserving fluorescent proteins and eliminating the need for hydrophilization when performing immunostaining.

1.2 Decolorization

Heme from blood and pigments derived from melanin are present in biological specimens and cause light absorption, which compromises tissue clearing. Decolorization of these pigments is an important step in tissue clearing, as even minute amounts of these pigments interfere with the acquisition of high-contrast images, especially 3D imaging with light-sheet microscopy. Heme, the most abundant pigment in biological samples, is associated with hemoglobin in blood and myoglobin in muscle. Ueda and colleagues reported that amino alcohols effectively decolorize heme [11, 12]. In addition, the latest CUBIC-L contains N-butyldiethanolamine [20]. A screening conducted by Tainaka and colleagues in 2018 showed that methylimidazole can also decolorize heme. Melanin, which is present in the eye and skin, can be decolorized with hydrogen peroxide [21–24]. The combination of CUBIC and hydrogen peroxide can clear mouse eyes and insects. However, strong oxidation using this method may cause tissue and antigenic damage and fade fluorescent proteins.

1.3 RI Matching

Suppressing light scattering is important for tissue clearing, especially Mie scattering caused by different RIs. Among biological materials, water typically has the lowest RI (~1.33), dried lipids have an intermediate RI (1.46–1.48), dried proteins and DNA have higher RIs (1.540–1.598 for dried proteins and ~1.58 for dried DNA), and hydroxyapatite extracted from bone tissue has the highest RI (1.600–1.604) [25–28]. Based on these known RIs, RI matching reagents are often prepared with an RI of ~1.50–1.60. Deformations, such as tissue shrinkage and swelling, should be considered when adjusting the RI. In general, hydrophobic clearing requires a higher RI due to tissue shrinkage by dehydration, whereas hydrophilic clearing often causes tissue swelling, allowing high transparency at lower RIs. The RI of the CUBIC-R + reagent (1.522), which is composed of antipyrine and nicotinamide, is lower than that of hydrophobic RI matching reagents like BABB. But high transparency can be achieved by swelling the specimens. The relationship between the density of the specimens and their RI is famously described by the Lorentz-Lorenz equation. However, in practice, tissues and RI matching reagents are often composed of complex mixtures. Therefore, it is necessary to adjust the expansion rate and the refractive index of the RI matching reagent while checking the final transparency of the tissues.

2 Whole-Brain Atlases

The brain is composed of various cellular circuits. A comprehensive analysis of the cellular circuits of the entire brain is a major challenge in neuroscience. The Allen Brain Atlas subdivides the adult brain into 900 neuroanatomical regions, representing the distribution of neurons, oligodendrocytes, astrocytes, microglia, and other major cells [29]. The Waxholm Space Atlas, the Franklin and Paxinos Atlas, the Brain/MINDS Marmoset Reference Atlas, and other atlases have also been constructed [30–33]. Applications using these atlases have been reported, including in situ hybridization and microarray combined with genomic expression data and neural analysis based on Adeno-Associated Virus (AAV) or rabies viruses [34–37]. In 2018, Murakami and colleagues reported a whole mouse brain atlas, including all cell coordinates, called the CUBIC-Atlas [38]. This atlas was created by expanding the brain volume 10× with a specially developed tissue-expandable CUBIC-X reagent. This technique was combined with a custom-made high-resolution light-sheet microscope to obtain fine images of an entire mouse brain with more than 100 million cells [34, 38]. Although image-based atlases have been used previously, the single-cell resolution atlas is expected to advance neuroscience by allowing more detailed analyses. Furthermore, since the CUBIC-Atlas uses spatial information of cell nuclei in the entire brain, this atlas is more detailed than image-based atlases and the data volume is smaller, which reduces storage and high computational costs for 3D biology studies. Regional information from the CUBIC-Atlas refers to the Allen Brain Atlas and is compatible with previous studies. The CUBIC-Atlas is integrated with CUBIC-Cloud (described below) and is available to global users on a cloud-based platform, and several cell types have already been annotated at the single-cell resolution [37].

3 Whole-Brain/Organ Staining Method

To extract information from intact tissues without sectioning, uniformly staining specific cells with nuclear staining and immunostaining deep into the tissue is required. However, penetration of stains into the interior during 3D staining is difficult. For nuclear staining, the use of low molecular weight dyes, such as PI and RedDot2, can facilitate diffusion into biological tissues under highly salt concentration [39]. For immunostaining, antibody penetration into the interior tissue is difficult due to strong binding to antigens on the tissue surface. To improve the permeability of antibodies, the salt concentration can be controlled, additional compounds can be added, the temperature can be increased, and the staining time can be increased. Several approaches can increase antibody penetration in thicker samples: small antibodies such as nanobody, urea or SDS to weaken antibody–tissue interactions, magnetohydrodynamic force, and increasing antibody volume [16, 39–43]. Although these methods have successfully increased the permeability of some antibodies, no general method has been established. In 2020, Susaki and colleagues described the CUBIC-HistVIsion staining protocol. In this protocol, fixed organs are an electrolyte gel that swells and shrinks reversibly, and the permeability of antibodies is tested using an artificial gelatin gel. Their results show that modulation of antibody–tissue interactions by chemical additives such as quadrol and urea can improve antibody permeability and that selection of appropriate temperature, ionic strength, detergent concentration, and antibody concentration is important. In addition, when immunostaining thick tissues, pre-conjugation of the antibodies can improve the penetration of thick tissues, as separate penetration of the primary and secondary antibodies is difficult. Thus, several methods of 3D immunostaining have been developed, but the methods still need improvements and require optimization for each antibody.

4 Whole-Organ Labeling with Fluorescent Proteins

The use of viruses or AAV is compatible with tissue clearing. Rabies virus for neuroprojection analysis and AAV-PHP.eB fluorescent protein, which can infect the whole brain, in combination with various promoters, allow quantification of any cell in the whole organ. Thus, organs stained with fluorescent proteins can also be immunostained using suitable tissue clearing and 3D immunostaining methods, and more information can be obtained.

5 Data Sharing Platform CUBIC-Cloud

Progress in tissue-clearing technology has facilitated the analysis of large tissues, such as the whole mouse brain, at the single-cell resolution. With these advances, a technique to efficiently utilize mapping data sets obtained from whole-mouse-brain imaging using light-sheet microscopy is needed. Mouse whole-brain mapping frameworks are being developed, including the CUBIC-Atlas. Mano and colleagues created CUBIC-Cloud (Fig. 2), a cloud-based tool for the quantification, visualization, and integration of mouse brain data [37]. The CUBIC-Cloud server automatically aligns uploaded brain data with the CUBIC-Atlas to create a user-specific database. CUBIC-Cloud also provides a graphic user interface for quantification work, enabling users to visualize uploaded brain data using the 3D Viewer. These data can be published in the CUBIC-Cloud public repository, allowing users around the world to view the data. CUBIC-Cloud is hosted at https://cubic-cloud.com. The following CUBIC-Cloud applications have been reported: (1) investigation of cell-type distribution in the brain, (2) quantification of brain Aβ plaque deposition in mouse models of Alzheimer's disease, (3) reconstruction of neuronal activity profiles during LPS-induced inflammation using cfos staining, and (4) whole-brain connectivity mapping using the rabies virus. As CUBIC-Cloud is a general-purpose analysis platform, any

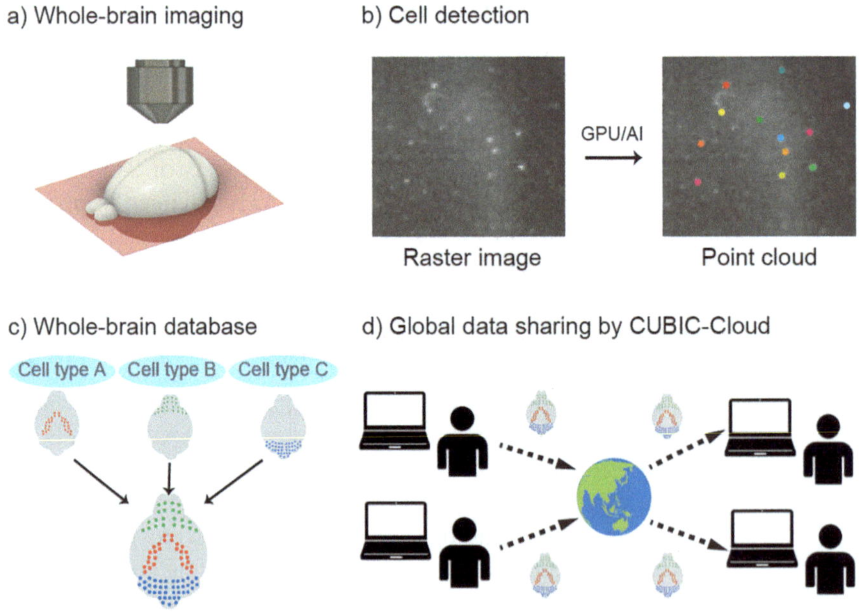

Fig. 2 Overview of the whole-brain analysis using CUBIC-Atlas and CUBIC-Cloud. Cleared mouse brains are imaged with LSFM (**a**), and cells of interest are detected and converted into a point cloud (**b**). The point clouds are integrated into CUBIC-Atlas (**c**) and shared with researchers with CUBIC-Cloud (**d**)

cell population can be analyzed if the cell information is mapped on the atlas. Thus, applications of CUBIC-Cloud are not limited to the above examples. The tool is currently used to analyze data from mouse brains, but the framework can be extended when tissue-clearing technology is adapted to rats and primates. Tools such as CUBIC-Cloud will become even more useful for collecting and analyzing brain data from a wide variety of organisms.

6 Example for Tissue Clearing and Imaging for Mouse Brain

The application of the tissue-clearing technique CUBIC enables the visualization of the activity of all neurons throughout the whole brain (Fig. 3). This process involves tissue clearing using CUBIC-L/R+ followed by three-dimensional immunostaining for c-Fos protein, a marker of neuronal activity, using the CUBIC-HV method [39]. The c-Fos signals are then imaged by light-sheet microscopy, and the CUBIC-Atlas can be used to facilitate quantitative evaluation of neural activity across all regions of the brain at the single-cell level. This method makes it possible to quantitatively analyze how neuronal activity in specific brain regions changes in response to pharmacological agents or external stimuli that affect the central nervous system. Comprehensive analysis of the effects of drug administration on the nervous system is expected to contribute to drug discovery. Furthermore, in the clinical field, it is expected to be applied to 3D pathological diagnosis.

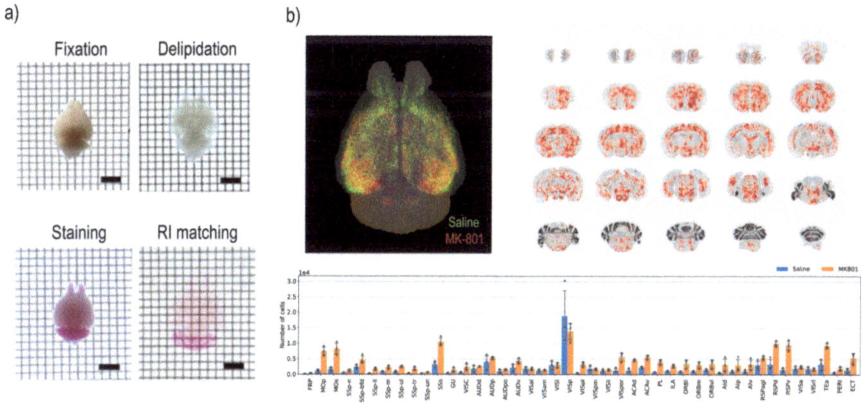

Fig. 3 Comprehensive analysis of whole brain activity using the CUBIC pipeline. (**a**) Tissue clearing and staining with CUBIC reagents. (**b**) Analysis of whole brain neural activity following administration of MK-801, an antagonist of the NMDA receptor. 3D view of c-Fos positive cells (top left). P-values for the density of c-Fos positive cells (top right). Bar graph showing the number of c-Fos positive cells in each brain region

7 Applications Using Tissue Clearing Beyond Neuroscience

Pathological diagnoses, including determination of cellular malignancy, degree of inflammation and fibrosis, and depth of tumor invasion, are based on the traditional examination of stained tissue on glass slides using optical microscopy. However, these traditional observation methods are limited to two-dimensional structures, and three-dimensional structures cannot be observed. In addition, only representative lesions are evaluated, and hidden lesions may be missed. To enable more efficient pathological diagnoses, 3D image acquisition from serial sections has been attempted; however, these methods are impractical due to the cost and labor requirements. Recent attempts have been made to combine tissue clearing and pathological diagnosis, including 3D quantification of the pancreas of streptozocin-induced diabetic mice, quantification of structures formed by neurites of autistic patients, and quantitative analysis of the distribution, structure, and spatial relationship of β-amyloid plaques with glial cells and blood vessels [12, 41]. Nojima and colleagues applied CUBIC to human organs to test the applicability of tissue clearing to clinical pathology. The 3D imaging of human lung and lymph node tissue clearing was possible, and the application of CUBIC increased the sensitivity of lymph node metastasis detection by 14.8% [18]. Thus, 3D pathology may improve the sensitivity of definitive diagnoses of cancer metastasis. Furthermore, Kubota and colleagues combined 3D pathology with machine learning in a mouse model of lung metastasis using human lung adenocarcinoma A549 cells; spatial information about the tumor microenvironment was successfully obtained [44, 45]. These reports indicate that tissue clearing and 3D imaging have the potential to extend the scope of pathological examination, and 3D pathology is a potential companion to AI. Thus, tissue-clearing technology and 3D imaging are expected to be the basis for the next generation of pathology.

Conflict of Interest

The authors have no conflicts of interest to declare that are relevant to the content of this chapter.

References

1. Lundvall H (1905) Weiteres über demonstration embryonaler Skelette. Anat Anz 27:520–523
2. Spalteholz W (1914) Über das Durchsichtigmachen von menschlichen und tierischen Präparaten. S. Hirzel
3. Bakutkin VV, M. IL, Semyonova TN, Tuchin VV, Kon IL (1995) Controlling of optical properties of sclera. SPIE 2393:137–141
4. Tuchin VV (1997) Light scattering of tissues. Physics 40(5):495–515

5. Tuchin VV, Maksimova IL, Zimnyakov DA, Kon IL, Mavlutov AH, Mishin AA (1997) Light propagation in tissues with controlled optical properties. J Biomed Opt 2(4): 401–417
6. Chung K et al (2013) Structural and molecular interrogation of intact biological systems. Nature 497(7449):332–337
7. Dodt HU et al (2007) Ultramicroscopy: three-dimensional visualization of neuronal networks in the whole mouse brain. Nat Methods 4(4): 331–336
8. Erturk A et al (2011) Three-dimensional imaging of the unsectioned adult spinal cord to assess axon regeneration and glial responses after injury. Nat Med 18(1):166–171
9. Erturk A, Lafkas D, Chalouni C (2014) Imaging cleared intact biological systems at a cellular level by 3DISCO. J Vis Exp (89)
10. Chi J et al (2018) Threedimensional adipose tissue imaging reveals regional variation in beige fat biogenesis and PRDM16-dependent sympathetic neurite density. Cell Metab 27(1): 226–236.e3
11. Susaki EA et al (2014) Whole-brain imaging with single-cell resolution using chemical cocktails and computational analysis. Cell 157(3): 726–739
12. Tainaka K et al (2014) Whole-body imaging with single-cell resolution by tissue decolorization. Cell 159(4):911–924
13. Tainaka K et al (2016) Chemical principles in tissue clearing and staining protocols for whole-body cell profiling. Annu Rev Cell Dev Biol 32:713–741
14. Hama H et al (2011) Scale: a chemical approach for fluorescence imaging and reconstruction of transparent mouse brain. Nat Neurosci 14(11):1481–1488
15. Ren Z et al (2021) CUBIC-plus: an optimized method for rapid tissue clearing and decolorization. Biochem Biophys Res Commun 568: 116–123
16. Murray E et al (2015) Simple, scalable proteomic imaging for high-dimensional profiling of intact systems. Cell 163(6):1500–1514
17. Inoue M et al (2019) Rapid chemical clearing of white matter in the post-mortem human brain by 1,2-hexanediol delipidation. Bioorg Med Chem Lett 29(15):1886–1890
18. Nojima S et al (2017) CUBIC pathology: three-dimensional imaging for pathological diagnosis. Sci Rep 7(1):9269
19. Zhao S et al (2020) Cellular and molecular probing of intact human organs. Cell 180(4): 796–812.e19
20. Tainaka K et al (2018) Chemical landscape for tissue clearing based on hydrophilic reagents. Cell Rep 24(8):2196–2210.e9
21. Kim SY, Assawachananont J (2016) A new method to visualize the intact subretina from retinal pigment epithelium to retinal tissue in whole mount of pigmented mouse eyes. Transl Vis Sci Technol 5(1):6
22. Futami K et al (2020) Application of hydrogen peroxide-melanin bleaching and fluorescent nuclear staining for whole-body clearing and imaging in fish. Fish Pathol 54(4):101–103
23. Kuroda M, Kuroda S (2020) Whole-body clearing of beetles by successive treatment with hydrogen peroxide and CUBIC reagents. Entomol Sci 23(3):311–315
24. Pende M et al (2020) A versatile depigmentation, clearing, and labeling method for exploring nervous system diversity. Sci Adv 6(22): eaba0365
25. Kienle DF et al (2014) Thickness and refractive index of DPPC and DPPE monolayers by multiple-beam interferometry. Anal Bioanal Chem 406(19):4725–4733
26. Alexey NB et al (2018) Measurement of tissue optical properties in the context of tissue optical clearing. J Biomed Opt 23(9):091416
27. Antonio A (1949) Quantitative researches on the optical properties of human bone. Nature 163(4146):604–604
28. Ascenzi A, Fabry C (1959) Technique for dissection and measurement of refractive index of osteones. J Biophys Biochem Cytol 6(1): 139–142
29. Dong HW (2008) The Allen reference atlas: a digital color brain atlas of the C57Bl/6J male mouse. Wiley
30. Papp EA et al (2014) Waxholm space atlas of the Sprague Dawley rat brain. NeuroImage 97: 374–386
31. Johnson GA et al (2010) Waxholm space: an image-based reference for coordinating mouse brain research. NeuroImage 53(2):365–372
32. Paxinos G, Franklin KBJ (2007) The mouse brain in stereotaxic coordinates. Academic Press, San Diego
33. Woodward A et al (2018) The Brain/MINDS 3D digital marmoset brain atlas. Sci Data 5: 180009
34. Matsumoto K et al (2019) Advanced CUBIC tissue clearing for whole-organ cell profiling. Nat Protoc 14(12):3506–3537
35. Zaeck L et al (2019) High-resolution 3D imaging of rabies virus infection in solvent-cleared brain tissue. J Vis Exp (146)

36. Roelofs TJM et al (2021) Optimization of whole-brain rabies virus tracing technology for small cell populations. Sci Rep 11(1):10400
37. Mano T et al (2021) CUBIC-Cloud provides an integrative computational framework toward community-driven whole-mouse-brain mapping. Cell Rep Methods 1(2):100038
38. Murakami TC et al (2018) A three-dimensional single-cell-resolution whole-brain atlas using CUBIC-X expansion microscopy and tissue clearing. Nat Neurosci 21(4): 625–637
39. Susaki EA et al (2020) versatile whole-organ/body staining and imaging based on electrolyte-gel properties of biological tissues. Nat Commun 11(1):1982
40. Fang T et al (2018) Nanobody immunostaining for correlated light and electron microscopy with preservation of ultrastructure. Nat Methods 15(12):1029–1032
41. Hama H et al (2015) ScaleS: an optical clearing palette for biological imaging. Nat Neurosci 18(10):1518–1529
42. Kim S-Y et al (2015) Stochastic electrotransport selectively enhances the transport of highly electromobile molecules. Proc Natl Acad Sci USA 112(46):E6274–E6283
43. Dwyer J et al (2021) Accelerated clearing and molecular labeling of biological tissues using magnetohydrodynamic force. Sci Rep 11(1): 16462
44. Kubota SI et al (2017) Whole-body profiling of cancer metastasis with single-cell resolution. Cell Rep 20(1):236–250
45. Kubota SI et al (2021) Whole-organ analysis of TGF-β-mediated remodelling of the tumour microenvironment by tissue clearing. Commun Biol 4(1):294

Part II

Pushing LSFM: Higher Resolution, Better Contrast, Larger Samples

Chapter 3

Advanced Tissue Clearing and Axially Swept Light-Sheet Microscopy for High-Resolution Imaging of Neural Systems

Hazel M. Borges, John Haug, Jinlong Lin, Samantha Calvache, Felix Y. Zhou, Xuemei Luo, Iliodora V. Pop, Zhenyu Zhong, Shuang Liang, Helen C. Lai, and Kevin M. Dean

Abstract

Imaging the peripheral and central nervous system requires high-resolution techniques capable of resolving fine axonal structures and synaptic connections across large, three-dimensional volumes. These tissues pose unique challenges due to their dense, heterogeneous composition and the intricate, branching architecture of nerve fibers, necessitating imaging approaches that combine deep tissue penetration, isotropic resolution, and minimal photodamage. This chapter outlines a robust protocol for tissue clearing and imaging using axially swept light-sheet microscopy (ASLM). We describe the optical principles of light-sheet fluorescence microscopy (LSFM), the challenges of imaging cleared tissues, and the critical trade-offs between field of view, resolution, and optical sectioning. By comparing the performance of two ASLM systems—bench-top mesoSPIM and a custom-designed cleared tissue (CT)-ASLM—we evaluate their capabilities in imaging complex neural and organ systems. Quantitative analyses of image sharpness, staining uniformity, and aggregate reduction serve as valuable metrics for optimizing tissue labeling, clearing, and microscope operation. This chapter provides practical guidelines for implementing LSFM in biological research, offering insights into optimizing tissue clearing protocols, imaging parameters, and downstream data analysis to advance the study of cellular and subcellular structures within native tissue contexts.

Key words Light-sheet fluorescence microscopy, Axially swept light-sheet microscopy, Tissue clearing, Quantitative analysis, Neural imaging

1 Introduction: High-Resolution Imaging of Neural Circuits

Proprioception, the sense of body and limb position, is mediated by specialized spinal circuits that relay sensory information from the periphery to the brain. Clarke's column (CC) neurons are a key spinal population sending hindlimb proprioception information ipsilaterally to the cerebellum [1–3]. CC axons travel several centimeters in the mouse from the spinal cord to the brain. However, their axon diameter is ~0.5 μm, making it challenging to follow

Ludovico Silvestri (ed.), *Light Sheet Microscopy*, Neuromethods, vol. 229, https://doi.org/10.1007/978-1-0716-5100-1_3,

individual axon trajectories that could inform their circuit connectivity and function. Likewise, as part of the autonomic nervous system, nerves associated with the hepatic portal vein are thought to regulate critical liver functions, such as controlling blood flow and influencing metabolic processes. Additionally, the liver's nervous system, comprising both afferent and efferent neurons, plays a key role in sensing changes in lipid, glucose, and metabolite levels and initiating appropriate physiological responses [4]. Despite these known functions, the exact roles and mechanisms of liver–nerve communication remain incompletely understood.

Proprioceptive and autonomic neural circuits, such as those in the spinal pathways and hepatic portal system, share the common challenge of being embedded in complex tissue environments, making their study reliant on advanced imaging techniques. To unravel the intricate cellular and subcellular structures of these neural systems—whether in the context of motor coordination or metabolic regulation—high-resolution methods capable of probing synaptic-scale dynamics within deep tissues are essential. Among the most employed approaches are two- and three-photon laser scanning microscopy, which permit the observation of biological structure and dynamics deep within living specimens [5]. Here, imaging is typically performed following the surgical exposure of the tissue of interest and/or introduction of an optical window; however, even with adaptive optics, the imaging depth is typically limited to ~1 mm for optically advantageous tissues [6]. Given their serial acquisition format, two- and three-photon laser scanning microscopy are slow, and their resulting anisotropic resolution obscures details along the optical axis of the microscope. More recently, advances in volume electron microscopy permit the observation of biological features at resolutions that approach molecular scales. These methods include focused ion beam scanning electron microscopy and transmission electron microscopy. In the former, superficial layers of a tissue are repetitively milled away and imaged to reconstruct a 3D volume [7]. In the latter, thinly sectioned tissues are iteratively loaded into high-speed imaging systems [8]. Nonetheless, both methods are especially time-consuming for large volumes and provide limited molecular specificity.

To effectively study neural circuits, imaging methods should not only be high-resolution but also rapid. Such approaches allow for the examination of numerous samples, thus capturing conserved features within the network that significantly contribute to function. Light-sheet fluorescence microscopy (LSFM), particularly variants optimized for high-resolution imaging, stands out in this regard. LSFM has revolutionized biological imaging by enabling high-speed, three-dimensional visualization of specimens with minimal photodamage. Here, the specimen is illuminated with a thin sheet of light, and fluorescence emission is collected orthogonally with a scientific camera. Fueled by recent advances in camera technologies, LSFMs now acquire vast volumes of high signal-to-noise data with rates up to gigavoxels per second. Consequently, LSFMs are exceptionally suited for high-replicate imaging of neurobiological processes, facilitating new insights into biological heterogeneity and disease mechanisms.

Despite the strengths of LSFM, it is often restricted to imaging only the most superficial layers of a tissue, typically less than one mean free scattering path [9]. This limitation arises because LSFM is highly sensitive to scattering and aberrations that affect both the delivery of illumination light to the specimen and the detection of fluorescence from the tissue, which compromises resolution and image quality at greater depths. To overcome this limitation, researchers often chemically clear the specimens, rendering it more transparent and thus increasing the mean free scattering path. However, tissue clearing becomes significantly more challenging as the numerical aperture (NA) of the imaging system increases, as higher NA optics are particularly sensitive to refractive index mismatches and optical inhomogeneities within the sample. In this chapter, we provide an overview of LSFMs that have been specifically optimized for cleared tissue imaging and outline an optimized clearing protocol that enables high optical transparency in both hard and soft tissues. Additionally, we demonstrate the imaging of diverse specimens on two different LSFMs and quantitatively evaluate image sharpness and contrast as a function of depth. By sharing these insights, we aim to democratize high-resolution cleared tissue imaging and open new opportunities for cell biology studies within complex, native tissue environments.

1.1 Optical Principles of LSFM

One of the fundamental design considerations of an LSFM is the combination of its illumination and detection objectives. These objectives must be arranged so that they fit orthogonally to each other at a 90° angle with their focal planes overlapping precisely (Fig. 1a). However, this configuration inherently involves a trade-off in NA. For instance, if the detection objective has a high NA, which provides superior resolution and sensitivity, the corresponding illumination objective must have a lower NA to fit geometrically, as the combined solid angle of both objectives cannot surpass 180° (Fig. 1b). Complicating matters further, objectives are typically optimized for operation at a single refractive index, which presents challenges for imaging cleared tissues, whose refractive indices can range significantly from 1.333 to 1.56 depending on the clearing technique used. To facilitate deep imaging into cleared tissue specimens, objectives should possess long working distances. However, this requirement often comes at the expense of NA, unless exceptionally large optics are employed. Due to these constraints, many light-sheet microscopes

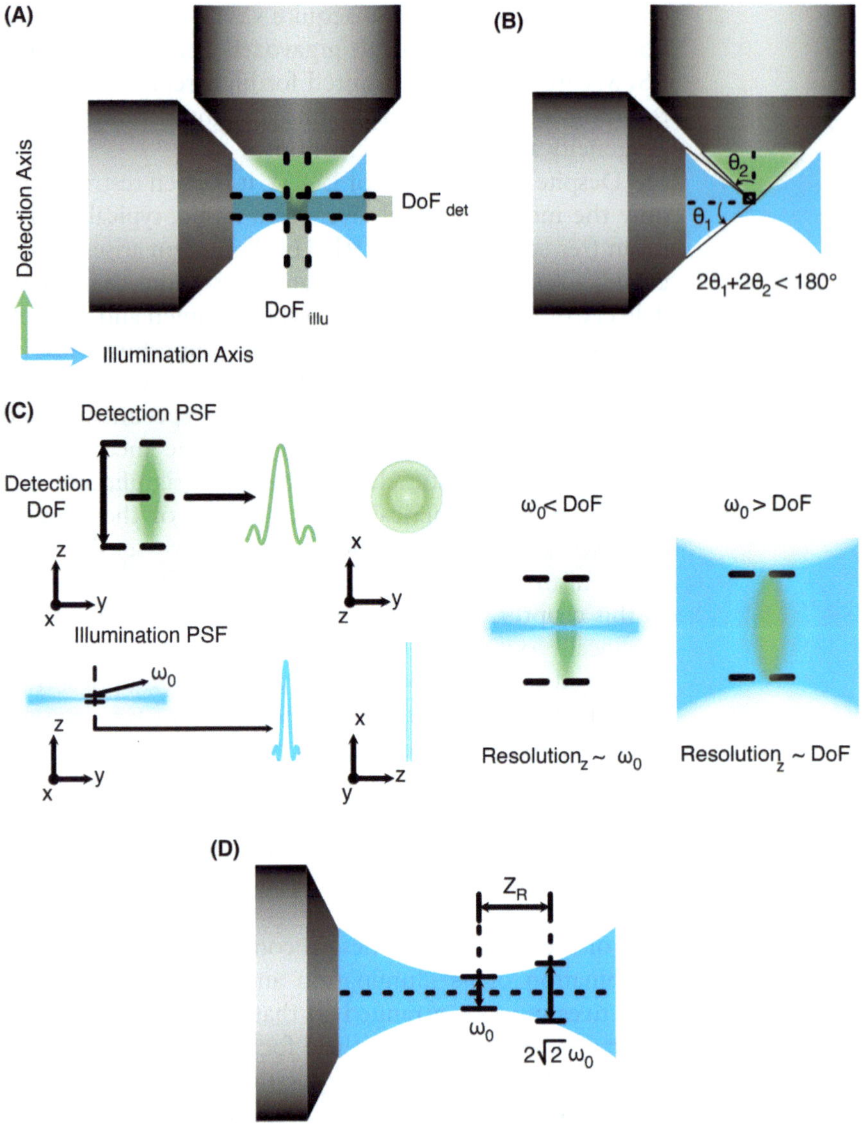

Fig. 1 (**a**) Schematic showing the ideal placement of the objective lenses in an orthogonal orientation, where the respective depth of focus for each objective is overlapping. (**b**) Schematic showing how the selection of the two objective lenses used in light-sheet configurations are limited by the cone of light angles associated with their respective NAs. (**c**) (First two rows) Schematic of an example detection and illumination point spread function in light-sheet microscopy. (Bottom row) Visual of the interplay between light-sheet width and detection depth of focus (DoF) for determining the resulting z-resolution of a light-sheet microscope. (**d**) Depiction of an example light-sheet, characterized by its width ω_0 and Rayleigh length Z_R

designed for cleared tissues use air illumination objectives and moderate NA detection objectives, sometimes augmented with a dipping cap to reduce spherical aberrations that become more pronounced with depth.

Under ideal, aberration-free and diffraction-limited imaging conditions, the lateral resolution in LSFM is determined solely by the emission wavelength of the fluorophore and the NA of the detection objective (Eq. 1). However, axial resolution is influenced by both the characteristics of the illuminating light-sheet and the detection optics. This interplay is captured by the cumulative point spread function (PSF) of the system, which is numerically defined as the product of the PSFs of both the illumination and detection optics (Eq. 2, Fig. 1c) [10, 11]. Because the cumulative PSF, and consequently the resolution, is highly sensitive to optical aberrations, the choice of objectives is particularly critical. Using any objective not specifically designed to operate at the refractive index of the specimen, or relying on air immersion objectives, can significantly degrade the microscope's resolution, especially at higher NAs. Addressing these challenges, specialized objectives for cleared tissue imaging have been developed which can function across a range of refractive indices 1.333–1.56 [12].

$$R_{XY} = \frac{\lambda_{\mathrm{em}}}{2\,\mathrm{NA}_{\mathrm{det}}} \quad (1)$$

$$R_z = \left(\frac{2\,\mathrm{NA}_{\mathrm{ex}}}{\lambda_{\mathrm{ex}}} + \frac{n(1-\cos\theta_{\mathrm{det}})}{\lambda_{\mathrm{em}}}\right)^{-1} \quad (2)$$

Another key property of LSFM is balancing the trade-off between field of view, axial resolution, and optical sectioning. Ideally, a LSFM based upon one-photon, linear absorption would achieve a large field of view while maintaining high axial resolution and superior optical sectioning. However, this balance is complex. When a tightly focused beam is used for illumination—where the beam waist is smaller than the depth of focus of the detection objective—the axial resolution approaches that of the beam waist (Fig. 1d). The drawback is that such tightly focused beams have a short Rayleigh length, which restricts the field of view that can be imaged effectively. Typically, the field of view of a LSFM is limited to the distance over which the dimensions of the illumination beam can be approximated as constant, or 2 Rayleigh lengths (Z_R, Fig. 1d). Conversely, illuminating the specimen with a beam that has a large Rayleigh length enables larger field-of-view imaging, but its larger beam waist causes the axial resolution of the system to approach the dimensions of the depth of focus of the detection objective. While this may increase the field of view, any fluorescence generated beyond the depth of focus of the detection objective will be detected as out-of-focus blur, reducing the optical sectioning capability of the microscope. This delicate interplay between field of view, axial resolution, and optical sectioning is fundamental to optimizing LSFM for various biological imaging needs.

1.2 Imaging Large, Chemically Cleared Tissues with LSFM

Despite these inherent trade-offs, LSFM systems optimized for imaging chemically cleared tissues have evolved rapidly. The earliest variants primarily used cylindrical lenses or air immersion objectives to illuminate the specimen, greatly limiting their resolution and ability to resolve fine detail in biological specimens [13]. However, with advances in optical engineering, a variety of innovative LSFM designs have emerged to expand the possibilities for cleared tissue imaging. For example, non-orthogonal scan or illumination mechanisms have been adopted to facilitate imaging of samples with lateral dimensions exceeding the working distance of both illumination and detection objectives. This approach is exemplified by dual inverted selective plane illumination microscope (diSPIM)-like and open-top LSFM configurations, which image samples from above or below, respectively, using a combination of orthogonal [14–16] or oblique [17, 18] illumination (Fig. 2a). Moreover, LSFMs leveraging multi-directional or near-infrared illumination have been deployed to reduce the attenuation of light from absorption and pushed the boundaries of imaging depth [19–21]. Multi-directional illumination effectively doubles the field of view by illuminating the specimen from both sides, reducing the path length that the illumination beam must travel through the tissue. In contrast, near-infrared illumination takes advantage of reduced scattering and absorption at longer wavelengths. These diverse approaches have enabled significant advancements in cleared tissue imaging, providing improved resolution, deeper penetration, and more uniform imaging across larger and more complex samples.

Despite these advances, only a handful of strategies attempt to navigate the fundamental trade-off between field of view and axial resolution. The most straightforward approach is to use a tightly focused beam and increase the amount of optical tiling necessary to image a large volume [22–24]. Optical tiling involves imaging multiple small fields of view with intentional overlap between them, which are then computationally fused to create a seamless, continuous dataset [25]. However, this method requires extensive computational post-processing, which reduces the overall throughput of the imaging system and complicates data handling. A more sophisticated approach involves the use of time delay integration (TDI) cameras, which synchronizes the transport of photoelectrons along a high-aspect-ratio camera register (e.g., 16 k × 256 pixels) with the movement of the specimen, effectively performing tiling in real-time [26, 27]. Unfortunately, TDI cameras are primarily designed for industrial applications and tend to have lower sensitivity, higher noise, and reduced bit depth compared to scientific imaging systems, limiting their utility in high-resolution biological imaging. Other strategies have included the use of propagation-invariant beams, such as Airy [28] and Bessel beams [29]. While these beams can maintain their focus over extended propagation lengths, their one-photon illumination implementations introduce

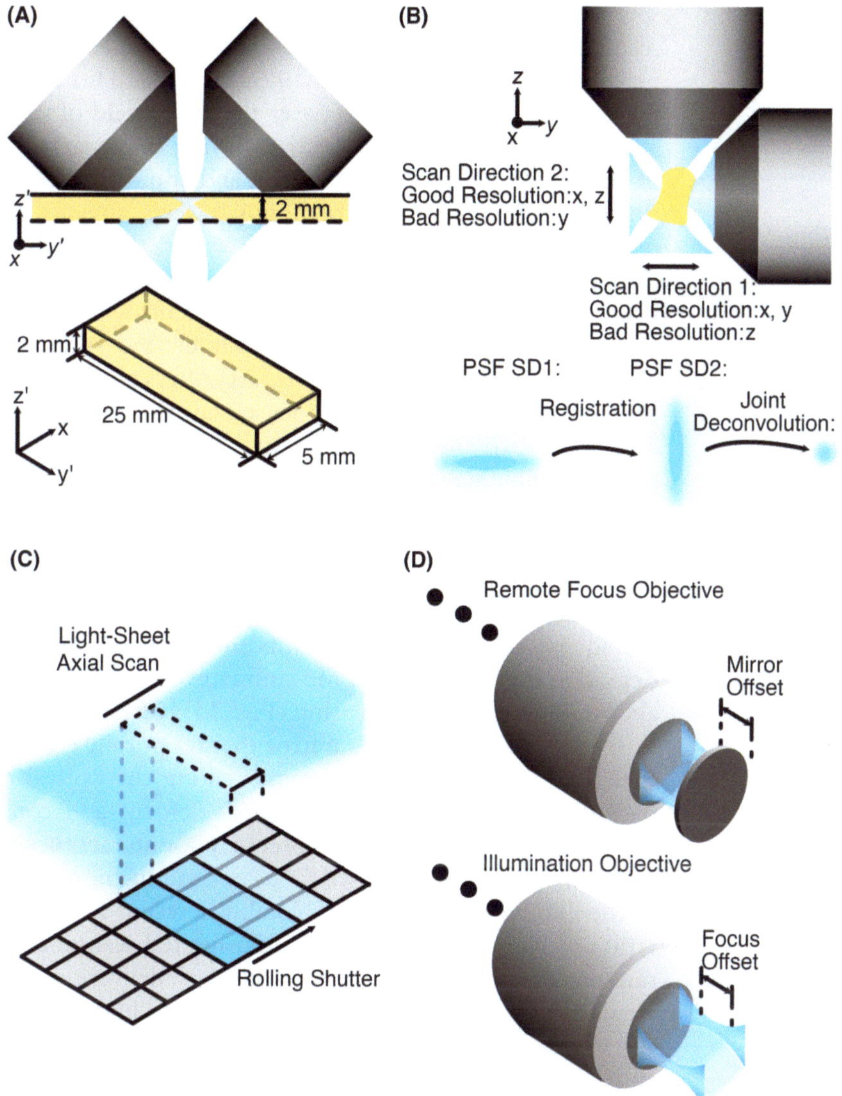

Fig. 2 (**a**) Example of a diSPIM-like configuration, where the potential imaging area can be extended indefinitely in the *x* and *y'* directions (here shown as 25 and 5 mm, respectively). (**b**) Concept of scanning directions for diSPIM configurations, where joint deconvolution approaches are implemented to utilize the best resolution associated with each scan direction. (**c**) Concept of ASLM, where the translation of a light-sheet across the field of view of a camera is synchronized with the camera's rolling shutter. (**d**) Depiction of how, in ASLM, the translation of a mirror placed in front of a remote focusing objective results in a corresponding translation of the light-sheet focus at the sample

complex side lobe structures that increase with beam length. These side lobes can generate unwanted fluorescence that degrades image quality, necessitating non-linear deconvolution to mitigate their effects. Even with such corrective measures, propagation-invariant beams often provide only marginal benefits over classical Gaussian beams [30–32].

An alternative approach to overcoming the limitations of axial resolution and field of view is to image the sample from orthogonal perspectives, as popularized by diSPIM [33]. This method uses a light sheet broad enough to cover the entire field of view, with the axial dimension of one perspective serving as the lateral dimension of another (Fig. 2b). By registering and applying joint deconvolution to the data from these orthogonal views, an isotropic imaging resolution can be achieved [33]. However, this approach necessitates the acquisition of multiple volumes and significant computational post-processing, adding to the already considerable data demands of imaging large, chemically cleared specimens. Axially swept light-sheet microscopy (ASLM) offers an alternative by synchronously scanning a tightly focused light sheet in its propagation direction with the rolling shutter on the camera (Fig. 2c) [34, 35]. This synchronization results in high axial resolution and optical sectioning across large fields of view. Higher NA versions of ASLM have demonstrated subcellular resolution in cleared tissues, achieving ~300 nm in chemically cleared specimens. In traditional high-NA systems, scanning the beam along its propagation path would typically introduce significant spherical aberrations that degrade the focus. ASLM mitigates this by incorporating aberration-free remote focusing (Fig. 2d) [36]. Here, the illumination light is directed onto a mirror that is optically conjugate to the specimen, and scanning this mirror along the optical axis of the remote focusing objective introduces defocus and compensatory aberrations. These aberrations precisely counteract those needed to maintain a diffraction-limited focus beyond the nominal focal plane of the illumination objective. Notably, a recent NA 0.7 implementation of ASLM achieved diffraction-limited scanning up to ±170 microns from the nominal depth of focus, delivering large fields of view with exceptional optical sectioning and high axial resolution.

Given this strategy, ASLM has found widespread adoption in cleared tissue imaging. In 2019, Voigt et al. introduced the mesoscale selective plane illumination microscope (mesoSPIM), which combined stop-shifted digital single-lens reflex (DSLR) lenses, and a macro zoom microscope to capture fields of view from 2.1 to 13.29 mm at high resolution [37]. More recently, a simplified version called the bench-top mesoSPIM (BT-mesoSPIM) was developed with fixed magnification objectives for a more compact and accessible design that maintained robust imaging performance [38]. For clinical specimens, an open-top variant of ASLM was developed that provided better contrast in densely labeled specimens [39]. Likewise, a recent version built in a diSPIM-like format implemented a mechanically sheared sample scanning format to achieve a large field-of-view imaging of clinical specimens while avoiding computational postprocessing of the data [40]. Alternative axial scanning techniques have also been explored, including spatially multiplexed light-sheets, the use of linear MEMS phased arrays [41], and structured illumination methods designed to suppress scattered light [42]. Collectively, these innovations highlight the continuous evolution of ASLM and its adaptations, each contributing to more effective, high-resolution, user-friendly [43], and versatile imaging of biological specimens. Through these developments, ASLM continues to provide researchers with a powerful tool for studying complex tissue structures at unprecedented detail.

1.3 Overview of Tissue Clearing Methods

Various tissue clearing methods have been developed to cater to different research needs, each with unique strengths and limitations. Broadly, these methods can be categorized into hyperhydrating techniques (e.g., CUBIC), organic solvent-based approaches (e.g., benzyl alcohol and benzyl benzoate), and hydrogel-based protocols (e.g., CLARITY and expansion microscopy) [44]. The fundamental aim of all these clearing strategies is to minimize light scattering caused by the heterogeneous composition of tissues, which results from the different refractive indices of water, lipids, proteins, etc., thereby facilitating the passage of light through the specimen. The choice of clearing method depends heavily on the research question and the specific properties of the tissue. For instance, while some protocols are compatible with fluorescent proteins, others necessitate immunofluorescence or the use of bio-orthogonal labeling techniques such as Halo and SNAP tags [45]. Although most tissue clearing methods have been developed and optimized for brain tissue, significant challenges remain when applying these protocols to other types of tissue, especially dense, epithelial-rich specimens. For a more detailed discussion on the variety of clearing methods available, we point the reader to several excellent reviews on the topic [44, 46].

One notable challenge as sample size increases is diffusion. Whether for delipidation or immunofluorescence labeling, the rate and extent of diffusion can become limiting. Diffusion is described by Fick's second law, and the time required for antibodies or reagents to penetrate a tissue is expected to grow quadratically with the distance they must travel. To address this, active clearing methods incorporating electrophoresis have been designed to propel reagents deeper into tissues, ensuring more uniform clearing and staining [47]. Optimizing tissue labeling and clearing is inherently multifactorial in nature, involving careful consideration of variables such as antibody diffusion, protein aggregation, reagent concentration, incubation periods, and the time and conditions for antibody removal after staining. Below, we present a detailed passive tissue clearing protocol that achieves consistent results and that we have tailored for specimens approximately 2 mm thick. We chose this thickness because it provided optically robust results across diverse tissue contexts, allowing for imaging through the entire tissue while maintaining diffraction-limited or near-diffraction-limited resolution. This protocol was adapted from the methods of Dodt et al. and Acar et al. [19, 48], which have been successfully applied to tissues such as the liver, lung, kidney, pancreas, intact bones, and brain, offering a practical solution for researchers seeking high optical transparency and effective immunolabeling in even the most challenging tissue types.

2 Materials

2.1 Tissue Clearing Reagents

- 4% Paraformaldehyde (PFA): prepared in PBS for tissue fixation.
- PBS with 0.02% sodium azide: used for washing tissues after fixation.
- 25% QUADROL: used for decolorization of soft and hard tissues.
- 0.5 M EDTA (pH 7.4): used for decalcification of hard tissues.
- Blocking buffer: 0.5% NP40, 10% DMSO, 0.5% Triton X-100, 5% serum (matching the host species of the secondary antibody), 1× PBS.
- Wash buffer: 0.5% NP40, 10% DMSO, 1× PBS.
- Primary and secondary antibody solutions: prepared in blocking buffer at desired concentrations.
- Nuclear stains: TO-PRO-3 (2 drops/mL) or SYTOX Green 488 (1 μM) in PBS.
- Methanol gradient: 25%, 50%, 75%, and 100% methanol in DiH_2O for tissue dehydration.
- Dichloromethane: used for delipidation.
- Benzyl alcohol and benzyl benzoate (BABB): 1:2 mixture used for refractive index matching, prepared by adding 5 g of activated aluminum oxide to 45 mL of BABB, rotating for 1 h at room temperature, and centrifuging at ~3000 rpm for 10 min to remove contaminants.
- Aluminum foil: used to protect samples from light.
- Eppendorf tubes (1.5 or 2.0 mL): used for tissue incubation.
- Ted Pella tissue slicer matrix: used for sectioning tissues into ~2 mm slices.
- Tube rotator: used for antibody incubation.
- MyFuge mini centrifuge: used for brief centrifugation of antibodies.

2.2 Animal Protocols

Details regarding the genetically modified mouse specimens used in this study have been previously reported [3]. All animal experiments were conducted in accordance with ethical guidelines and were approved by the Institutional Animal Care and Use Committee (IACUC) at UT Southwestern Medical Center.

Table 1
Basic optical parameters for BT-mesoSPIM and CT-ASLM

	BT-mesoSPIM	CT-ASLM
Detection objective	XLFLUOR4X/340	ASI 54-12-8
Numerical aperture ($\eta = 1.333$)	0.28	0.64
Numerical aperture ($\eta = 1.56$)	0.28	0.75
Tube lens focal length (mm)	200	300
Magnification ($\eta = 1.333$)	4×	32.81×
Magnification ($\eta = 1.56$)	4×	38.39×
Field of view (mm, $\eta = 1.333$)	3.3 × 3.3	0.41 × 0.41
Field of view (mm, $\eta = 1.56$)	3.3 × 3.3	0.35 × 0.35

2.3 Imaging Systems

All imaging is performed with a BT-mesoSPIM or an NA 0.7 CT-ASLM. Table 1 summarizes the basic optical parameters of each system, while more comprehensive details, including complete parts lists, are available in their respective publications [12, 35, 37, 38]. The BT-mesoSPIM system is equipped with a Nikon AF-S 50 mm f/1.4 G air-immersion objective for illumination and an Olympus XLFLUOR4X/340 4× long-working-distance air-immersion detection objective. The sample is placed within a 50-mm optical cuvette (Hellma), enabling large-scale imaging at moderate numerical apertures. In contrast, CT-ASLM systems are equipped with multi-immersion objectives that are in direct contact with the solvent, and their effective focal length, numerical aperture, and magnification vary depending upon the refractive index of the medium. These systems, paired with a standard 2048 × 2048 scientific CMOS camera, provide a wide range of imaging fields of view, from 350 × 350 microns to 3.3 × 3.3 mm. The numerical apertures span from 0.28 to 0.75, encompassing a solid cone angle of approximately 34.5–72° at a refractive index of 1.56. This variety of numerical apertures offers a valuable case study to illustrate how NA can affect imaging performance, resolution, and optical sectioning in the context of cleared tissue imaging. Each system is run with *navigate*, a user-friendly, open-source, Python-based LSFM control software that enables users to automate LSFM workflows and provides real-time image-based feedback [49].

3 Methods

3.1 A Robust Tissue Clearing Protocol for Hard and Soft Tissues

Prior to tissue harvesting, mice are deeply anesthetized and perfused with phosphate-buffered saline (PBS) using a peristaltic pump with a volume flow rate of 20 mL/min until the liver turns white, indicative of complete blood removal. Thereafter, paraformaldehyde solution at a final concentration of 4% in PBS is pumped throughout the mouse until the tail becomes firm. Tissues are then surgically harvested and fixed overnight in 4% PFA at 4 °C with continuous shaking. Following fixation, tissues are thoroughly washed three times in PBS with 0.02% sodium azide for 2 h at 4 °C to remove any residual PFA. Complete perfusion is essential to ensure the removal of blood and is sufficiently achieved when the liver appears white. Perfusion not only reduces tissue autofluorescence but also greatly improves its optical transparency. The tissues are stored in this PBS solution at 4 °C until further processing. Prior to clearing, tissues undergo pre-treatment based on their type: soft tissues require decolorization, while hard tissues need both decolorization and decalcification. Typically, soft tissues are sectioned into ~2 mm slices using a tissue slicer matrix (Ted Pella, Inc. #15020) and placed in 2-mL Eppendorf tubes with 25% QUADROL for decolorization. For hard tissues, decalcification is achieved using 0.5 M EDTA (pH 7.4) at 37 °C with constant shaking over 5 days, with daily solution changes. Completion of decalcification is evident when the bone is spongy and easy to cut [50]. EDTA is rinsed off with three quick washes with deionized water, and the sample is then decolorized with a 25% QUADROL solution at 37 °C. For both hard and soft tissues, the QUADROL solution is regularly changed until the supernatant becomes colorless, which can be monitored quantitatively with a spectrophotometer at ~400 and ~600 nm.

After pre-treatment, the tissues are permeabilized and blocked in a buffer solution containing 0.5% NP40, 10% DMSO, 0.5% Triton X-100, 5% serum, and 1× PBS overnight at room temperature. The choice of serum for blocking should match the host species of the secondary antibody to prevent non-specific binding and reduce background during imaging. Importantly, to avoid protein aggregation, primary and secondary antibodies are briefly centrifuged and retrieved from the top of the stock solution before dilution in the blocking buffer. Tissues are incubated in the primary antibody solution for 72 h for soft tissues and 96 h for hard tissues in a tube rotator at room temperature, followed by three 2-h washing steps in a buffer consisting of 0.5% NP40, 10% DMSO, and 1× PBS and then post-fixed with 4% PFA overnight. If multiple antigens are being stained, it is important to carefully select different primary antibody hosts, and unconjugated primary antibodies are preferred as they tend to penetrate the specimen more deeply.

Post-fixation of primary antibodies is found to greatly reduce the number of fluorescent aggregates in the specimen, presumably by preventing transient dissociation events of the primary antibody from the antigen in the presence of a high concentration of secondary antibodies. After another wash in PBS, tissues are stained with secondary antibodies for 72 h and subsequently washed over 2 days. Nuclei are stained using TO-PRO-3 or SYTOX Green at a concentration of ~2 drops/mL of 1–5 micromolar, respectively, for 48 h, with PBS washes in between. Although the manufacturer recommends phosphate-free buffers for staining DNA, we found that PBS provides superior penetration into the specimen, presumably due to the dye competitively binding with free phosphate instead of DNA. It is critical to protect samples from light throughout the process to prevent photobleaching. Careful selection and preparation of antibodies compatible with the organic-based clearing protocol are necessary, especially for multicolor imaging, to avoid cross-reactions and ensure deep tissue penetration.

Following immunostaining, tissues are dehydrated in a methanol/DiH_2O gradient (25%, 50%, 75%, 100%) at room temperature for 1 h per concentration. This step allows water to be gradually replaced, preserving tissue integrity. Dehydration is completed with two 45-min incubations in 100% methanol. Tissues are then iteratively incubated with fresh dichloromethane for 45 min until the tissue sinks, indicating an increase in specimen density and thorough removal of lipids. For 2-mm-thick tissue specimens, two cycles are typically sufficient. This step is essential for eliminating molecules that contribute to light scattering. Finally, tissues are transferred to a 1:2 mixture of benzyl alcohol and benzyl benzoate (BABB) for refractive index matching, achieving a homogenous index of 1.56. To remove peroxide contaminants, prepared BABB is treated with activated aluminum oxide for an hour at room temperature with constant shaking. To remove the aluminum oxide pellets, the BABB solution is centrifuged at ~3000 rpm for 10 min and then decanted into a clean container. To match the refractive index of the tissues, they are first incubated with constant rotation in BABB for 15 min. Thereafter, the BABB solution is refreshed, and the tissue is incubated overnight. Prior to imaging, the BABB solution is refreshed once more for another hour. All BABB incubation steps are performed at room temperature, and the sample can remain in BABB indefinitely until imaging.

Optimizing staining parameters such as antibody dilution or concentration, staining buffer composition, incubation duration, pre-treatment steps, antibody post-fixation, and wash times is crucial after selecting the appropriate primary and secondary antibodies. We often employ dyes like Alexa Fluor or CF dyes, which are known for providing high-contrast imaging across a variety of solvents. However, this level of performance is not universal; for example, TO-PRO-3 appears significantly brighter in aqueous contexts, suggesting that its fluorescence is sensitive to environmental conditions. The chosen antibody concentration and incubation period must balance achieving a strong signal-to-noise ratio while minimizing non-specific binding and the potential formation of aggregates. Although this protocol has proven effective across various tissue types, the incubation times for each step may need adjustment based on the tissue's size. This protocol not only ensures high transparency and uniformity within the specimen but also minimizes optical inhomogeneities, facilitating optimal imaging quality in cleared tissue LSFM systems.

3.2 Optimization of ASLM Imaging Parameters

To obtain high-quality data in ASLM, it is essential to carefully optimize both the alignment of the microscope and the synchronization of the light-sheet with the rolling shutter of the camera. Proper alignment ensures that the light-sheet intersects the depth of focus of the detection objective, while synchronization is crucial for generating the thin, optical sections that enable high-resolution imaging throughout a large field of view. To adjust the synchronization of the light-sheet, use the "Waveform Parameters" button in navigate (highlighted in red, Fig. 3a). This will open the Waveform Parameter Settings popup window (Fig. 3b), allowing independent control of the waveform parameters for each laser. These adjustments enable fine-tuning of the light-sheet's synchronization with the rolling shutter of the camera.

The BT-mesoSPIM is a digitally scanned variant of ASLM that creates a virtual sheet of light by sweeping a focused Gaussian beam laterally across the camera during the exposure. Simultaneously, the beam is scanned in its propagation direction using an electrotunable lens. For optimal alignment, the lateral sweep of the Gaussian beam can be disabled by adjusting the Galvo amplitude in the Waveform Parameter Settings popup. The amplitude and offset of the Remote Focus settings are then iteratively adjusted to achieve a narrow line beam. Alignment can be performed on the solvent alone to prevent unnecessary photobleaching of the specimen; we recommend moving the specimen vertically out of the illumination beam during this process. This procedure should be repeated for every laser intended for use to ensure optimal alignment across all excitation wavelengths, which can be selected in the Channels tab of the main graphical user interface (GUI). If a thin laser line cannot be established, consider adjusting the readout direction of the light-sheet mode in the Camera Settings tab (e.g., switching between top-to-bottom or bottom-to-top) to ensure proper synchronization between the laser and camera. For a detailed protocol, see Figure 9 in [35]. Once the optimized remote focusing settings have been identified, increase the amplitude of the Galvo such that it illuminates the entire field of view.

Aligning light-sheet microscopes that use a cylindrical lens to form the light-sheet, such as CT-ASLM, is more challenging

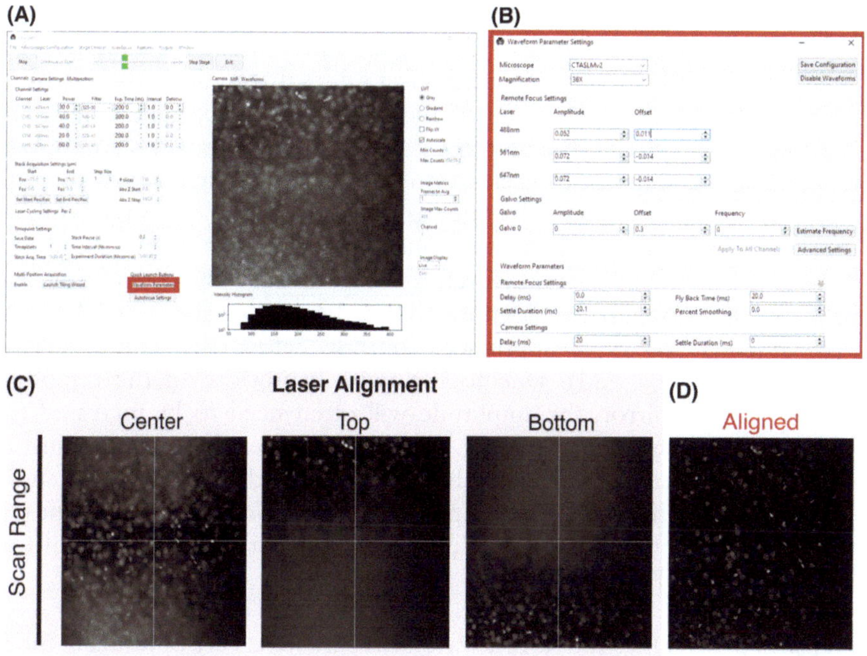

Fig. 3 (**a**) Navigate microscope control software. A blurry image is shown in the camera FOV before optimizing the waveform parameters settings. (**b**) Offset and amplitude should be modified for each laser before starting an image acquisition. Disable the waveform parameters (red) and adjust the remote focusing offset along with the focus position of the detection objective until the laser appears sharp and in focus at the center of the camera's field of view. (**c**) The offset should be adjusted to position the beam at the top and the bottom of the camera field of view. The difference in the offsets should be used as a starting point for setting the amplitude of the remote focusing beam. (**d**) The amplitude, offset, and focus should be iteratively adjusted until the entire image is sharp and uniform across the field of view, ensuring optimal imaging parameters

compared to digitally scanned systems. Here, for alignment optimization, it is easiest to work with densely labeled samples, such as SYTOX Green-stained nuclei. Alignment on the solvent alone is not feasible. The first step is to select the appropriate excitation laser and emission filter in the Channels tab. Next, ensure that the light-sheet is positioned within the detection objective's depth of focus (Fig. 3c, left). To achieve this, ensure the camera is operating in its normal mode, not the light-sheet mode, which can be controlled in the main GUI under the Camera Settings tab. Disable the scanning of the light-sheet in its propagation direction by pressing the "Disable Waveforms" button in the Waveform Parameters popup window. Adjust the focus of the detection objective so that a narrow strip of in-focus bright fluorescence is evident (Fig. 3c). Regions before and after the narrow strip should appear blurry, representing the converging and diverging portions of the laser focus, each generating out-of-focus blur. This adjustment can be performed manually or by using the focus stage, if one is configured within navigate.

Once the laser focus is optimized, the remote focusing amplitude and offset values can be adjusted. A good strategy is to identify the remote focusing offset values that position the laser focus at the top and bottom of the camera (Fig. 3c, middle and right). The difference between these offset values provides an estimate of the amplitude necessary to scan the light-sheet across the full field of view. Input this value as the amplitude in the Waveform Parameters popup as a starting condition for the laser being used. Next, select the "Enable Waveforms" button in the Waveform Parameters popup and incrementally adjust both the amplitude and offset values to optimize contrast across the entire field of view (Fig. 3d). Due to the finite bandwidth of the remote focusing mirror, the amplitude will often need to be increased relative to the initial estimate. For final alignment, iteratively adjust the focus of the detection objective along with the amplitude and offset of the remote focusing settings. Once properly aligned, the entire field of view will exhibit high contrast with no out-of-focus blur (Fig. 3d).

3.3 Acquisition of a Z-Stack

Once the ASLM waveform parameters are optimized, the next step is to select the regions of interest within the tissue for imaging. Begin by navigating to the Stage Control tab, where you can adjust the position of the stages. For convenience, tabs can be popped out by right-clicking on them, allowing easier access while performing adjustments. To set the Z-stack acquisition range, move the Z stage to the desired starting position and click the Set Start Pos/Foc button in the Channels tab. Then, move the Z stage to the desired ending position and click the Set End Pos/Foc button. In cases where a refractive index mismatch remains in the sample, a small focus shift may occur between the start and end positions. If this is observed, adjust the focus at each position before setting the start and end positions. This ensures that navigate will perform a focus ramp through the tissue during imaging, maintaining sharp focus throughout the stack. Note that a motorized focus axis configured with navigate is required for this feature. Additionally, an autofocus wizard is available in navigate, allowing the identification of the best focal plane to be automated, further streamlining the imaging setup process. The step size of the Z-stack is configured in the Channels tab and should be set to achieve Nyquist sampling, which is approximately half the axial resolution of the microscope. For CT-ASLM systems with NA 0.7 objectives, the step size is typically set between 160 and 200 nm to maintain isotropic resolution. In contrast, for BT-mesoSPIM, which operates at lower numerical apertures, a step size of approximately 1 micron is sufficient to capture the necessary resolution.

3.4 Automated Tiling of Large Image Volumes

When the desired imaging region exceeds the field of view, tiling becomes necessary to capture the entire volume. To initiate this process, select the Launch Tiling Wizard button in the Channels tab of navigate. Identify the start and end positions of the sample in the *X*, *Y*, and *Z* dimensions, and configure a focus ramp if needed. Specify the desired overlap between adjacent fields of view; this parameter depends on the sample's labeling density. For densely labeled samples, minimal overlap (~10%) is often sufficient, while sparse samples may require greater overlap to ensure continuity between tiles. Once these parameters are set, click the Populate Multi-Position Table to generate the tiling sequence. When saving the data, it is recommended to use an N5, H5, or Zarr format to enable immediate visualization with the BigDataViewer and facilitate seamless stitching with the BigStitcher plugin in FIJI [25, 51]. BigStitcher can computationally fuse overlapping tiles into a continuous dataset, providing a high-quality reconstruction of large image volumes.

3.5 Quantitative Evaluation of Labeling and Clearing Efficacy

Quantitative image analysis can be realized to evaluate the efficacy of the sample preparation and imaging performance. For example, the Shannon entropy of the discrete cosine transform (DCTS) can be used to measure image contrast as a function of depth to determine the sample clearing efficacy (Fig. 4a, b) [52]. While the DCTS metric is influenced by factors such as tissue architecture and the specific microscope used, it serves as an internally consistent measure of image sharpness. By analyzing the rate at which the DCTS decays, the transparency of the specimen under tissue preparation or imaging conditions—such as with and without delipidation, or for a given wavelength and numerical aperture—can be evaluated. For instance, with an NA 0.7 lens on the CT-ASLM, we observe a rapid initial decay in DCTS, suggesting that imaging remains more optimal at superficial tissue layers. Beyond transparency, bright aggregates caused by non-specific antibody binding pose a common challenge in tissue clearing. By deploying 3D spot detectors, such as the Laplacian of Gaussian, staining conditions can be systematically compared, as demonstrated in Fig. 4c [53]. This analysis highlights the efficacy of post-fixation in reducing aggregates, likely by stabilizing antibody–antigen interactions and preventing large complex formation in the presence of secondary antibodies. Although not shown here, additional informative metrics exist, including Fourier-based image correlation for assessing resolution and analysis of defocus as a function of depth to evaluate refractive index mismatch [54]. Collectively, such quantitative metrics provide valuable guidance for optimizing clearing and imaging parameters, ensuring the acquisition of high-quality data that is well-suited for downstream analyses.

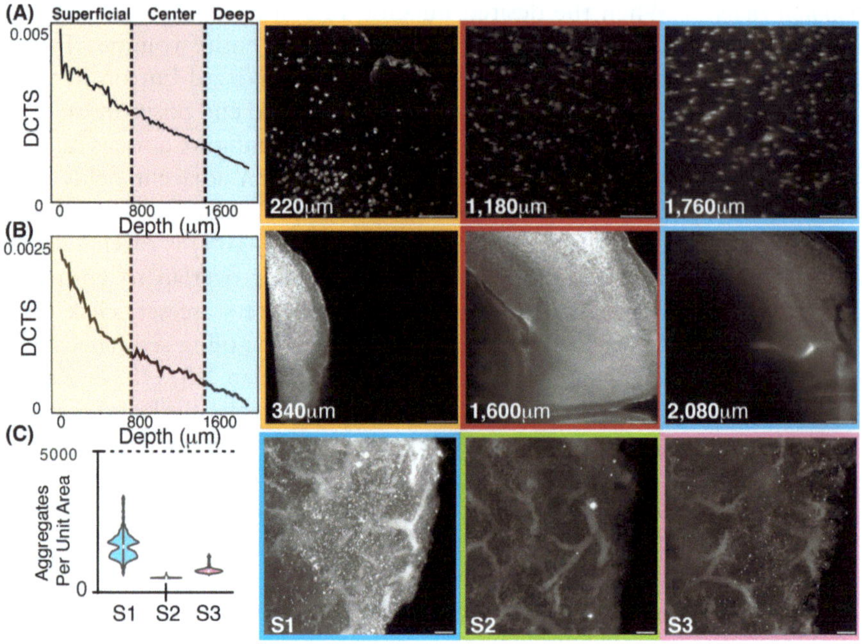

Fig. 4 (**a**) Image sharpness as a function of depth for CT-ASLM and (**b**) BT-mesoSPIM. Image sharpness is quantified using the Shannon entropy of the discrete cosine transform (DCTS). Columns 2, 3, and 4 display representative images from different depths, as indicated by the dashed lines in the first column. Cells are labeled with SYTOX Green. Scale bars represent 50 microns for (**a**) and 500 microns for (**b**). (**c**) Quantification of aggregates using a 3D Laplacian of Gaussian spot detector under different staining conditions: S1 (no post-fixation), S2 (post-fixation after primary staining), and S3 (no primary antibody). The number of spots is normalized to the tissue volume. Scale bars, 50 microns

3.6 High-Resolution Imaging of Autonomic Nerves with ASLM

To demonstrate the capabilities of CT-ASLM, we imaged autonomic nerves associated with the hepatic portal vein, a key component of the hepatic nervous plexus. These nerves integrate sympathetic fibers from the celiac plexus and parasympathetic fibers from the vagus nerve, playing critical roles in regulating liver functions such as controlling portal vein blood flow and influencing metabolic processes. Additionally, the liver's nervous system, comprising afferent and efferent neurons, is essential for sensing changes in lipid, glucose, and metabolite levels and initiating appropriate physiological responses. To visualize these nerves, we stained for Beta-Tubulin III (TUBB3) and imaged the chemically cleared tissue using BT-mesoSPIM in an automated tiling format with navigate. Figure 5a shows a volume rendering of the entire specimen, revealing a dense network of nerves propagating from right to left as the portal vein branches into hepatic sinusoids. High-resolution imaging of a local region with CT-ASLM (Fig. 5b) further highlights the complexity of these nerve bundles. Closer inspection reveals intricate branching patterns and bulbous,

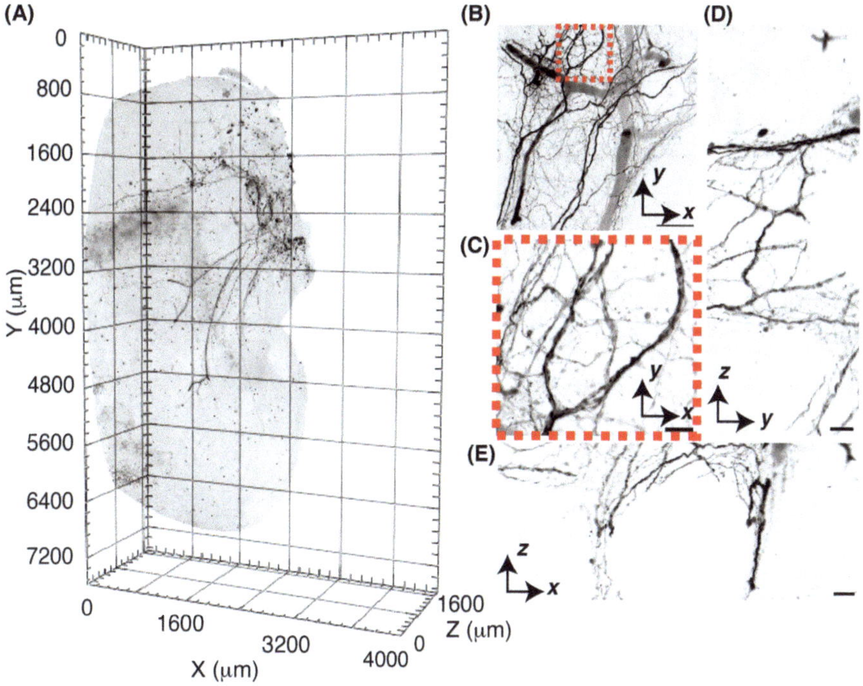

Fig. 5 (**a**) Volume rendering of an intact, chemically cleared liver lobule stained for Beta-Tubulin III (TUBB3) and imaged using the BT-mesoSPIM. Neurons are visible traversing the portal vein from right to left. (**b**) Local high-resolution imaging with CT-ASLM, showing a maximum intensity projection of a 342 × 342 × 200 μm region. (**c**) Dashed region from (**b**) displayed as a maximum intensity projection in the XY plane. (**d**) Corresponding maximum intensity projection in the YZ and (**e**) XZ planes

synapse-like structures visible in all spatial dimensions (Fig. 5c–e), underscoring the power of ASLM for detailed structural analysis of autonomic nerves in cleared tissue contexts.

3.7 High-Resolution Spinocerebellar Imaging

CC neurons are integral to direct spinocerebellar pathways, transmitting detailed proprioceptive inputs for motor coordination. Anatomically, CC neurons transmit proprioceptive signals through extensive mossy fiber projections that can terminate bilaterally in the cerebellar cortex without significant collateralization in the spinal cord or medulla. It remains poorly understood how distinct axonal terminations of CC neurons in the cerebellum translate into functional differences in motor coordination and proprioceptive processing. To evaluate these structures, we employed our tissue clearing method to render an intact mouse brain and spine optically transparent. The original specimen spans 70 mm, extending from the anterior of the brain to the lumbar vertebrae, as shown in Fig. 6a. CC neurons were sparsely labeled, enabling the visualization of their axonal projections over tens of millimeters with a BT-mesoSPIM (Fig. 6b, c). CC neuron axons can be seen extending ipsilaterally toward the cerebellar cortex (data not shown).

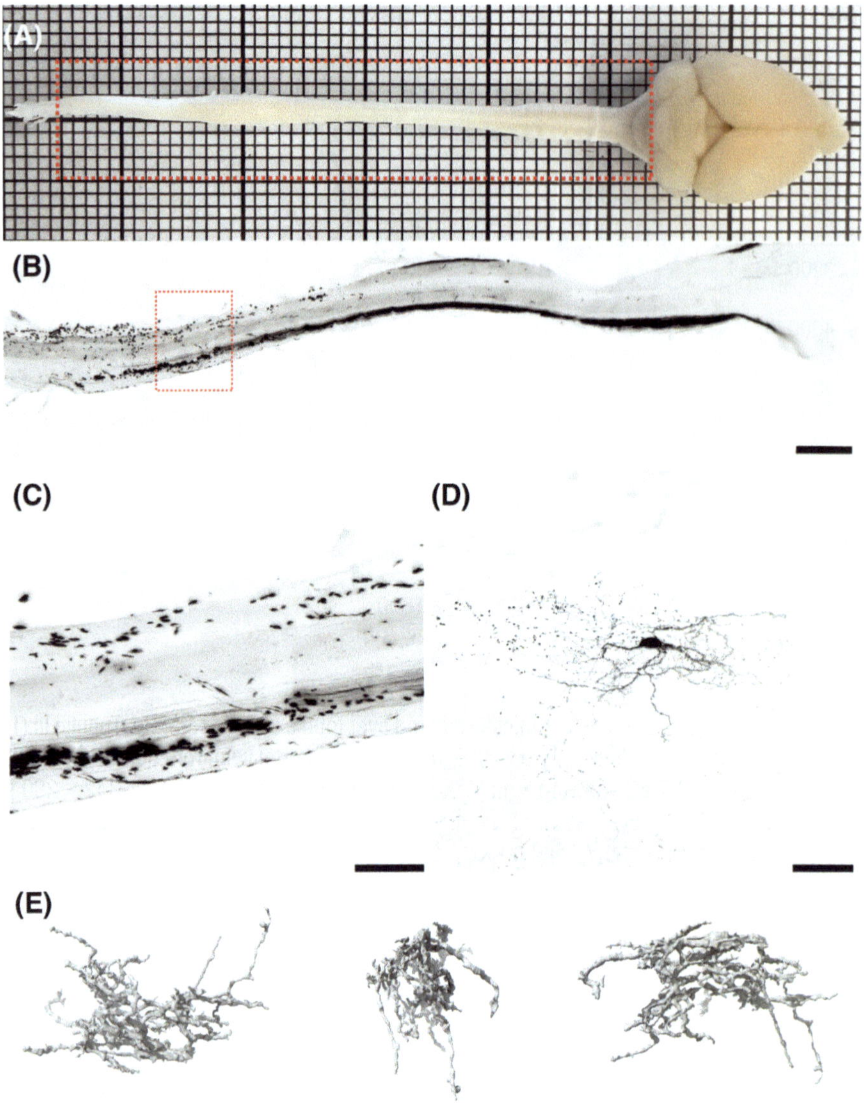

Fig. 6 (**a**) Intact mouse brain and spinal cord before optical clearing. Grid size (1 mm). Imaging region of the sample is shown in red rectangle. (**b**) A maximum intensity projection of a BABB cleared spinal cord with low-resolution imaging in BT mesoSPIM. Scale Bar, 2 mm. (**c**) Visualization of spinal projection neurons from (**b**). Scale bar, 500 μm. (**d**) A maximum intensity projection of a single neuron imaged in CT-ASLM with an isotropic resolution. Scale bar, 40 μm. (**e**) 3D volume rendering of segmented neurons, visualized with ChimeraX

Local, high-resolution imaging with CT-ASLM further resolved the dense, highly arborized, and synaptic structures of CC neurons in the spinal column (Fig. 6d). To morphologically analyze the neuron, we performed binary Otsu segmentation of the image after applying Meijering filtering to enhance vessel-like features, then specifically removing voxels belonging to non-neuron background cells which were detected using u-Segment3D [12], a

Python-based segmentation package. u-Segment3D segments the 3D data across the XY, XZ, and YZ planes using a pretrained 2D neural network segmentation model (here, Cellpose) [55, 56] and combining the independent 2D segmented stacks into one consensus 3D segmentation. This is done by using a distance transform to transform 2D cell masks into 2D gradients to assemble 3D gradients which are used identify individual 3D cells by spatial proximity clustering. While we used the default diffusion distance transform in this analysis to detect and remove the ellipsoidal non-neuron cells, u-Segment3D provides alternative transforms and tunable parameters that could be applied to tailor the segmentation to segment cells with more complex morphologies and optimization for specific imaging conditions or properties.

Following segmentation, the data were exported as a surface file and visualized in ChimeraX, a powerful rendering tool that uses ray tracing to generate 3D models with enhanced depth perception (Fig. 6e) [57]. Three perspectives of the segmented neuron are presented, showcasing its complex branching and synaptic architecture. This approach highlights the utility of combining tissue clearing, high-resolution light-sheet microscopy, and advanced computational tools to unravel the intricate organization of neural pathways, providing a foundation for understanding their functional implications in proprioceptive and motor signal integration.

4 Conclusions

In this chapter, we have outlined advanced methods for imaging cleared tissues using ASLM, highlighting its capabilities for high-resolution visualization of neural circuits and complex tissue structures. By integrating optimized tissue clearing protocols with quantitative analysis and sophisticated imaging systems, we demonstrated the potential of ASLM to resolve cellular and subcellular features in large, intact specimens. These approaches not only enable deeper insights into the structure and function of neural and autonomic systems but also provide a framework for studying diverse biological processes in their native tissue environments. We hope this chapter serves as a resource to facilitate the adoption of these techniques and inspire future advancements in tissue imaging and analysis.

Acknowledgements

This research was supported by funding from the NIH National Institute of General Medical Sciences (grant RM1GM145399 to K. M.D.) and the NIH National Cancer Institute (grant U54CA268072 to K.M.D.). Research in the labs of Zhong and

Liang was funded by the National Institute of Health (R35GM142654 to Z.Z. and R01DK133283 to S.L.) and by the Cancer Prevention and Research Institute of Texas (CPRIT) (RP230261 to Z.Z. and RP240183 to S.L.). ZZ is a CPRIT Scholar and supported by a CPRIT New Investigator Recruitment Award (RR180014). I.V.P. and H.C.L. were supported by funding from NIH NINDS R21NS128203.

Conflict of Interest The authors have no conflicts of interest to declare that are relevant to the content of this chapter.

References

1. Oscarsson O (1965) Functional organization of the spino- and cuneocerebellar tracts. Physiol Rev 45:495–522. https://doi.org/10.1152/physrev.1965.45.3.495
2. Espinosa F, Pop IV, Lai HC (2024) Electrophysiological properties of proprioception-related neurons in the intermediate thoracolumbar spinal cord. eNeuro 11(4). https://doi.org/10.1523/ENEURO.0331-23.2024
3. Pop IV, Espinosa F, Blevins CJ, Okafor PC, Ogujiofor OW, Goyal M, Mona B, Landy MA, Dean KM, Gurumurthy CB, Lai HC (2022) Structure of long-range direct and indirect spinocerebellar pathways as well as local spinal circuits mediating proprioception. J Neurosci 42(4):581–600. https://doi.org/10.1523/JNEUROSCI.2157-20.2021
4. Jensen KJ, Alpini G, Glaser S (2013) Hepatic nervous system and neurobiology of the liver. Compr Physiol 3(2):655–665. https://doi.org/10.1002/cphy.c120018
5. Ouzounov DG, Wang T, Wang M, Feng DD, Horton NG, Cruz-Hernandez JC, Cheng YT, Reimer J, Tolias AS, Nishimura N, Xu C (2017) In vivo three-photon imaging of activity of GCaMP6-labeled neurons deep in intact mouse brain. Nat Methods 14(4):388–390. https://doi.org/10.1038/nmeth.4183
6. Wang K, Milkie DE, Saxena A, Engerer P, Misgeld T, Bronner ME, Mumm J, Betzig E (2014) Rapid adaptive optical recovery of optimal resolution over large volumes. Nat Methods 11(6):625–628. https://doi.org/10.1038/nmeth.2925
7. Xu CS, Pang S, Shtengel G, Muller A, Ritter AT, Hoffman HK, Takemura SY, Lu Z, Pasolli HA, Iyer N, Chung J, Bennett D, Weigel AV, Freeman M, van Engelenburg SB, Walther TC, Farese RV Jr, Lippincott-Schwartz J, Mellman I, Solimena M, Hess HF (2021) An open-access volume electron microscopy atlas of whole cells and tissues. Nature 599(7883):147–151. https://doi.org/10.1038/s41586-021-03992-4
8. Schlegel P, Yin Y, Bates AS, Dorkenwald S, Eichler K, Brooks P, Han DS, Gkantia M, Dos Santos M, Munnelly EJ, Badalamente G, Serratosa Capdevila L, Sane VA, Fragniere AMC, Kiassat L, Pleijzier MW, Sturner T, Tamimi IFM, Dunne CR, Salgarella I, Javier A, Fang S, Perlman E, Kazimiers T, Jagannathan SR, Matsliah A, Sterling AR, Yu SC, McKellar CE, FlyWire C, Costa M, Seung HS, Murthy M, Hartenstein V, Bock DD, Jefferis G (2024) Whole-brain annotation and multi-connectome cell typing of Drosophila. Nature 634(8032):139–152. https://doi.org/10.1038/s41586-024-07686-5
9. Ntziachristos V (2010) Going deeper than microscopy: the optical imaging frontier in biology. Nat Methods 7(8):603–614. https://doi.org/10.1038/nmeth.1483
10. Remacha E, Friedrich L, Vermot J, Fahrbach FO (2020) How to define and optimize axial resolution in light-sheet microscopy: a simulation-based approach. Biomed Opt Express 11(1):8–26. https://doi.org/10.1364/BOE.11.000008
11. Gao L, Shao L, Chen BC, Betzig E (2014) 3D live fluorescence imaging of cellular dynamics using Bessel beam plane illumination microscopy. Nat Protoc 9(5):1083–1101. https://doi.org/10.1038/nprot.2014.087
12. Chakraborty T, Driscoll MK, Jeffery E, Murphy MM, Roudot P, Chang B-J, Vora S, Wong WM, Nielson CD, Zhang H, Zhemkov V, Hiremath C, De La Cruz ED, Yi Y, Bezprozvanny I, Zhao H, Tomer R, Heintzmann R, Meeks JP, Marciano DK, Morrison SJ, Danuser G, Dean KM, Fiolka R (2019) Light-sheet microscopy of cleared tissues with isotropic, subcellular resolution. Nat

Methods 16(11):1109–1113. https://doi.org/10.1038/s41592-019-0615-4
13. Voie AH, Burns DH, Spelman FA (1993) Orthogonal-plane fluorescence optical sectioning: three-dimensional imaging of macroscopic biological specimens. J Microsc 170(Pt 3): 229–236. https://doi.org/10.1111/j.1365-2818.1993.tb03346.x
14. Glaser AK, Reder NP, Chen Y, McCarty EF, Yin C, Wei L, Wang Y, True LD, Liu JTC (2017) Light-sheet microscopy for slide-free non-destructive pathology of large clinical specimens. Nat Biomed Eng 1(7). https://doi.org/10.1038/s41551-017-0084
15. Glaser AK, Reder NP, Chen Y, Yin C, Wei L, Kang S, Barner LA, Xie W, McCarty EF, Mao C, Halpern AR, Stoltzfus CR, Daniels JS, Gerner MY, Nicovich PR, Vaughan JC, True LD, Liu JTC (2019) Multi-immersion open-top light-sheet microscope for high-throughput imaging of cleared tissues. Nat Commun 10(1). https://doi.org/10.1038/s41467-019-10534-0
16. Schueth A, Hildebrand S, Samarska I, Sengupta S, Kiessling A, Herrler A, Zur Hausen A, Capalbo M, Roebroeck A (2023) Efficient 3D light-sheet imaging of very large-scale optically cleared human brain and prostate tissue samples. Commun Biol 6(1):170. https://doi.org/10.1038/s42003-023-04536-4
17. Migliori B, Datta MS, Dupre C, Apak MC, Asano S, Gao R, Boyden ES, Hermanson O, Yuste R, Tomer R (2018) Light sheet theta microscopy for rapid high-resolution imaging of large biological samples. BMC Biol 16(1). https://doi.org/10.1186/s12915-018-0521-8
18. Glaser AK, Bishop KW, Barner LA, Susaki EA, Kubota SI, Gao G, Serafin RB, Balaram P, Turschak E, Nicovich PR, Lai H, Lucas LAG, Yi Y, Nichols EK, Huang H, Reder NP, Wilson JJ, Sivakumar R, Shamskhou E, Stoltzfus CR, Wei X, Hempton AK, Pende M, Murawala P, Dodt HU, Imaizumi T, Shendure J, Beliveau BJ, Gerner MY, Xin L, Zhao H, True LD, Reid RC, Chandrashekar J, Ueda HR, Svoboda K, Liu JTC (2022) A hybrid open-top light-sheet microscope for versatile multi-scale imaging of cleared tissues. Nat Methods 19(5):613–619. https://doi.org/10.1038/s41592-022-01468-5
19. Dodt H-U, Leischner U, Schierloh A, Jährling N, Mauch CP, Deininger K, Deussing JM, Eder M, Zieglgänsberger W, Becker K (2007) Ultramicroscopy: three-dimensional visualization of neuronal networks in the whole mouse brain. Nat Methods 4(4): 331–336. https://doi.org/10.1038/nmeth1036
20. Stefaniuk M, Gualda EJ, Pawlowska M, Legutko D, Matryba P, Koza P, Konopka W, Owczarek D, Wawrzyniak M, Loza-Alvarez P, Kaczmarek L (2016) Light-sheet microscopy imaging of a whole cleared rat brain with Thy1-GFP transgene. Sci Rep 6:28209. https://doi.org/10.1038/srep28209
21. Wang F, Wan H, Ma Z, Zhong Y, Sun Q, Tian Y, Qu L, Du H, Zhang M, Li L, Ma H, Luo J, Liang Y, Li WJ, Hong G, Liu L, Dai H (2019) Light-sheet microscopy in the near-infrared II window. Nat Methods 16(6): 545–552. https://doi.org/10.1038/s41592-019-0398-7
22. Buytaert JA, Dirckx JJ (2007) Design and quantitative resolution measurements of an optical virtual sectioning three-dimensional imaging technique for biomedical specimens, featuring two-micrometer slicing resolution. J Biomed Opt 12(1):014039. https://doi.org/10.1117/1.2671712
23. Chen Y, Li X, Zhang D, Wang C, Feng R, Li X, Wen Y, Xu H, Zhang XS, Yang X, Chen Y, Feng Y, Zhou B, Chen BC, Lei K, Cai S, Jia JM, Gao L (2020) A versatile tiling light sheet microscope for imaging of cleared tissues. Cell Rep 33(5):108349. https://doi.org/10.1016/j.celrep.2020.108349
24. Wang D, Jin Y, Feng R, Chen Y, Gao L (2019) Tiling light sheet selective plane illumination microscopy using discontinuous light sheets. Opt Express 27(23):34472–34483. https://doi.org/10.1364/OE.27.034472
25. Horl D, Rojas Rusak F, Preusser F, Tillberg P, Randel N, Chhetri RK, Cardona A, Keller PJ, Harz H, Leonhardt H, Treier M, Preibisch S (2019) BigStitcher: reconstructing high-resolution image datasets of cleared and expanded samples. Nat Methods 16(9): 870–874. https://doi.org/10.1038/s41592-019-0501-0
26. Schacht P, Johnson SB, Santi PA (2010) Implementation of a continuous scanning procedure and a line scan camera for thin-sheet laser imaging microscopy. Biomed Opt Express 1(2): 598–609. https://doi.org/10.1364/BOE.1.000598/
27. Tang L, Wang J, Ding J, Sun J, Chen X-j, Shen Q, Song R, Cao P, Gong R, Xu F, Ge W-p, Sun W, Zhao H, Wu J (2024) Curved light sheet microscopy for centimeter-scale cleared tissues imaging. Nat Photonics. https://doi.org/10.1101/2024.04.28.591483
28. Nylk J, McCluskey K, Aggarwal S, Tello JA, Dholakia K (2016) Enhancement of image

quality and imaging depth with Airy light-sheet microscopy in cleared and non-cleared neural tissue. Biomed Opt Express 7(10):4021–4033. https://doi.org/10.1364/BOE.7.004021

29. Gao R, Asano SM, Upadhyayula S, Pisarev I, Milkie DE, Liu TL, Singh V, Graves A, Huynh GH, Zhao Y, Bogovic J, Colonell J, Ott CM, Zugates C, Tappan S, Rodriguez A, Mosaliganti KR, Sheu SH, Pasolli HA, Pang S, Xu CS, Megason SG, Hess H, Lippincott-Schwartz J, Hantman A, Rubin GM, Kirchhausen T, Saalfeld S, Aso Y, Boyden ES, Betzig E (2019) Cortical column and whole-brain imaging with molecular contrast and nanoscale resolution. Science 363(6424). https://doi.org/10.1126/science.aau8302
30. Chang B-J, Dean KM, Fiolka R (2020) Systematic and quantitative comparison of lattice and Gaussian light-sheets. Opt Express 28(18). https://doi.org/10.1364/oe.400164
31. Liu G, Ruan X, Milkie DE, Gorlitz F, Mueller M, Hercule W, Killilea A, Betzig E, Upadhyayula S (2023) Characterization, comparison, and optimization of lattice light sheets. Sci Adv 9(13):eade6623. https://doi.org/10.1126/sciadv.ade6623
32. Shi Y, Daugird TA, Legant WR (2022) A quantitative analysis of various patterns applied in lattice light sheet microscopy. Nat Commun 13(1):4607. https://doi.org/10.1038/s41467-022-32341-w
33. Wu Y, Wawrzusin P, Senseney J, Fischer RS, Christensen R, Santella A, York AG, Winter PW, Waterman CM, Bao Z, Colon-Ramos DA, McAuliffe M, Shroff H (2013) Spatially isotropic four-dimensional imaging with dual-view plane illumination microscopy. Nat Biotechnol 31(11):1032–1038. https://doi.org/10.1038/nbt.2713
34. Dean KM, Roudot P, Welf ES, Danuser G, Fiolka R (2015) Deconvolution-free subcellular imaging with axially swept light sheet microscopy. Biophys J 108(12):2807–2815. https://doi.org/10.1016/j.bpj.2015.05.013
35. Dean KM, Chakraborty T, Daetwyler S, Lin J, Garrelts G, M'Saad O, Mekbib HT, Voigt FF, Schaettin M, Stoeckli ET, Helmchen F, Bewersdorf J, Fiolka R (2022) Isotropic imaging across spatial scales with axially swept light-sheet microscopy. Nat Protoc. https://doi.org/10.1038/s41596-022-00706-6
36. Botcherby EJ, Smith CW, Kohl MM, Debarre D, Booth MJ, Juskaitis R, Paulsen O, Wilson T (2012) Aberration-free three-dimensional multiphoton imaging of neuronal activity at kHz rates. Proc Natl Acad Sci USA 109(8):2919–2924. https://doi.org/10.1073/pnas.1111662109
37. Voigt FF, Kirschenbaum D, Platonova E, Pages S, Campbell RAA, Kastli R, Schaettin M, Egolf L, van der Bourg A, Bethge P, Haenraets K, Frezel N, Topilko T, Perin P, Hillier D, Hildebrand S, Schueth A, Roebroeck A, Roska B, Stoeckli ET, Pizzala R, Renier N, Zeilhofer HU, Karayannis T, Ziegler U, Batti L, Holtmaat A, Luscher C, Aguzzi A, Helmchen F (2019) The mesoSPIM initiative: open-source light-sheet microscopes for imaging cleared tissue. Nat Methods 16(11):1105–1108. https://doi.org/10.1038/s41592-019-0554-0
38. Vladimirov N, Voigt FF, Naert T, Araujo GR, Cai R, Reuss AM, Zhao S, Schmid P, Hildebrand S, Schaettin M, Groos D, Mateos JM, Bethge P, Yamamoto T, Aerne V, Roebroeck A, Erturk A, Aguzzi A, Ziegler U, Stoeckli E, Baudis L, Lienkamp SS, Helmchen F (2024) Benchtop mesoSPIM: a next-generation open-source light-sheet microscope for cleared samples. Nat Commun 15(1):2679. https://doi.org/10.1038/s41467-024-46770-2
39. Bishop KW, Erion Barner LA, Baraznenok E, Lan L, Poudel C, Brenes D, Serafin RB, True LD, Vaughan JC, Glaser AK, Liu JTC (2024) Axially swept open-top light-sheet microscopy for densely labeled clinical specimens. Opt Lett 49(13):3794–3797. https://doi.org/10.1364/OL.521591
40. Lin J, Mehra D, Marin Z, Wang X, Borges HM, Shen Q, Galecki S, Haug J, Abbott DH, Dean KM (2024) Mechanically sheared axially swept light-sheet microscopy. Biomed Opt Express 15(9):5314–5327. https://doi.org/10.1364/BOE.526145
41. Landry J, Hamann S, Solgaard O (2020) High-speed axially swept light sheet microscopy using a linear MEMS phased array for isotropic resolution. J Biomed Opt 25(10). https://doi.org/10.1117/1.JBO.25.10.106504
42. Frantz D, Wright CJ, Schaser AJ, Kirik D, Kristensson E, Berrocal E (2024) Image post-processing for SILMAS: structured illumination light sheet microscopy with axial sweeping. Biomed Opt Express 15(8):4943–4962. https://doi.org/10.1364/BOE.531210
43. Liu Y, Rollins AM, Jenkins MW (2021) CompassLSM: axially swept light-sheet microscopy made simple. Biomed Opt Express 12(10):6571–6589. https://doi.org/10.1364/BOE.440292
44. Richardson DS, Guan W, Matsumoto K, Pan C, Chung K, Erturk A, Ueda HR, Lichtman JW (2021) Tissue clearing. Nat Rev Methods Primers 1(1). https://doi.org/10.1038/s43586-021-00080-9

45. Kohl J, Ng J, Cachero S, Ciabatti E, Dolan M-J, Sutcliffe B, Tozer A, Ruehle S, Krueger D, Frechter S, Branco T, Tripodi M, Jefferis GSXE (2014) Ultrafast tissue staining with chemical tags. Proc Natl Acad Sci 111(36). https://doi.org/10.1073/pnas.1411087111
46. Weiss KR, Voigt FF, Shepherd DP, Huisken J (2021) Tutorial: practical considerations for tissue clearing and imaging. Nat Protoc 16(6):2732–2748. https://doi.org/10.1038/s41596-021-00502-8
47. Kim SY, Cho JH, Murray E, Bakh N, Choi H, Ohn K, Ruelas L, Hubbert A, McCue M, Vassallo SL, Keller PJ, Chung K (2015) Stochastic electrotransport selectively enhances the transport of highly electromobile molecules. Proc Natl Acad Sci USA 112(46):E6274–E6283. https://doi.org/10.1073/pnas.1510133112
48. Acar M, Kocherlakota KS, Murphy MM, Peyer JG, Oguro H, Inra CN, Jaiyeola C, Zhao Z, Luby-Phelps K, Morrison SJ (2015) Deep imaging of bone marrow shows non-dividing stem cells are mainly perisinusoidal. Nature 526(7571):126–130. https://doi.org/10.1038/nature15250
49. Marin Z, Wang X, Collison DW, McFadden C, Lin J, Borges HM, Chen B, Mehra D, Shen Q, Galecki S, Daetwyler S, Sheppard SJ, Thien P, Porter BA, Conzen SD, Shepherd DP, Fiolka R, Dean KM (2024) Navigate: an open-source platform for smart light-sheet microscopy. Nat Methods. https://doi.org/10.1038/s41592-024-02413-4
50. Jing D, Zhang S, Luo W, Gao X, Men Y, Ma C, Liu X, Yi Y, Bugde A, Zhou BO, Zhao Z, Yuan Q, Feng JQ, Gao L, Ge WP, Zhao H (2018) Tissue clearing of both hard and soft tissue organs with the PEGASOS method. Cell Res 28(8):803–818. https://doi.org/10.1038/s41422-018-0049-z
51. Pietzsch T, Saalfeld S, Preibisch S, Tomancak P (2015) BigDataViewer: visualization and processing for large image data sets. Nat Methods 12(6):481–483. https://doi.org/10.1038/nmeth.3392
52. Royer LA, Lemon WC, Chhetri RK, Wan Y, Coleman M, Myers EW, Keller PJ (2016) Adaptive light-sheet microscopy for long-term, high-resolution imaging in living organisms. Nat Biotechnol 34(12):1267–1278. https://doi.org/10.1038/nbt.3708
53. Tinevez JY, Perry N, Schindelin J, Hoopes GM, Reynolds GD, Laplantine E, Bednarek SY, Shorte SL, Eliceiri KW (2017) TrackMate: an open and extensible platform for single-particle tracking. Methods 115:80–90. https://doi.org/10.1016/j.ymeth.2016.09.016
54. Descloux A, Grussmayer KS, Radenovic A (2019) Parameter-free image resolution estimation based on decorrelation analysis. Nat Methods 16(9):918–924. https://doi.org/10.1038/s41592-019-0515-7
55. Meijering E, Jacob M, Sarria JC, Steiner P, Hirling H, Unser M (2004) Design and validation of a tool for neurite tracing and analysis in fluorescence microscopy images. Cytometry A 58(2):167–176. https://doi.org/10.1002/cyto.a.20022
56. Stringer C, Wang T, Michaelos M, Pachitariu M (2021) Cellpose: a generalist algorithm for cellular segmentation. Nat Methods 18(1):100–106. https://doi.org/10.1038/s41592-020-01018-x
57. Meng EC, Goddard TD, Pettersen EF, Couch GS, Pearson ZJ, Morris JH, Ferrin TE (2023) UCSF ChimeraX: tools for structure building and analysis. Protein Sci 32(11):e4792. https://doi.org/10.1002/pro.4792

Chapter 4

Light-Sheet Fluorescence Microscopy of Expanded Neuronal Tissue

Juan Eduardo Rodriguez-Gatica, Martin Karl Schwarz, and Ulrich Kubitscheck

Abstract

Light-sheet microscopy allows rapid mesoscopic visualization of extended brain specimens. Combining this approach with tissue clearing and expansion is particularly fruitful. Here, we present a workflow for neuronal tissue clearing, expansion, and imaging that enables detailed mesoscopic mapping of neuronal connections with the ability to zoom in and out of selected neuronal circuits with virtual "super-resolution."

Key words Expansion microscopy, Fluorescence labeling, Autofluorescent proteins, Sample dimensions, Stitching, Connectivity, Clearing

1 Introduction

The development of light microscopic techniques with high spatial resolution and molecular contrast has been a key driver in the study of biological structures. Neuronal circuits, in particular, have been a focus of intense research in recent years, with new techniques allowing particular circuit elements to be labeled and made optically accessible. Trans-synaptic tracer viruses have also been successfully used to determine synaptic connections over long distances and, notably, can be used to unambiguously determine a synaptic connection [1]. These developments have greatly benefited from the ability to image spatially extended brain slices or even complete brains, which has been made possible through the combination of tissue clearing and light-sheet fluorescence microscopy [2]. Despite these technical advances, detailed mesoscopic mapping of neuronal connections that also allows "zooming" in and out of selected neuronal circuits with "super-resolution" has not been possible until recently. This can now be achieved through an additional physical expansion of the tissue before light-sheet imaging

Ludovico Silvestri (ed.), *Light Sheet Microscopy*, Neuromethods, vol. 229, https://doi.org/10.1007/978-1-0716-5100-1_4,

[3]. The biochemical process of expansion increases the effective optical resolution and—as a welcome side effect—also leads to a very effective clearing of the tissue. Samples can be kept in an aqueous buffer, allowing for the straightforward use of water immersion or water-dipping lenses for imaging.

Numerous groups have successfully used high-resolution imaging of extended, expanded, and cleared brain samples [4–10]. As a consequence of that, there are currently many different expansion techniques available, each with its pros and cons (reviewed by [11]). We present here the workflow to expand, clear, and investigate especially large samples of neuronal tissue with a thickness of up to 2 mm for expansion factors of 1.5–4. This workflow is based on recently published and extensively tested protocols [4, 6, 8, 12]. We describe the various steps of sample preparation starting with a paraformaldehyde-fixed mouse brain tissue section. We explain how the samples are labeled by immunofluorescence (Subheading 3.1) and additionally stained with a cell nucleus marker to obtain a general overview of the structure of the sample (Subheading 3.2) and how to perform the actual expansion process (Subheading 3.3). In Subheading 3.4, we describe the key steps of imaging such large and extended specimens and indicate the first computational post-processing steps of the data.

2 Materials

2.1 Buffers and Solutions

Phosphate-buffered saline (PBS): 136.89 mM NaCl, 10 mM $Na_2HPO_4{\cdot}2H_2O$, 1.98 mM KH_2PO_4, 2.68 mM KCl.

Washing buffer: PBS.

Permeabilization buffer: PBS, 0.5% w/v 3-[(3-cholami- dopropyl) dimethylammonio]-1-propanesulfonate (CHAPS).

Blocking buffer: 1× PBS, 5% normal goat serum, 0.3% Triton X-100, 0.02% sodium azide.

Digestion buffer: 50 mM Tris–HCl, 1 mM EDTA, 0.5% Triton X-100, 0.8 M guanidine HCl, and 16 U/mL of proteinase K; pH 8.0.

PBS-T: PBS containing 0.3% Triton X-100.

Linker solution: 1 M methylacrylic acid-NHS (MA-NHS) in DMSO. Store as 10 μL aliquots in −20 °C.

Monomer solution: Mix 2.25 mL of 19 g/50 mL sodium acrylate, 0.625 mL of 40% acrylamide (w/v), 0.75 mL of 1 g/50 mL N, N′-methylenebisacrylamide, 4 mL of 14.6 g/50 mL sodium chloride, 1 mL of 10× PBS, and 0.775 mL Milli Q water. Store as 600 μL aliquots in −20 °C.

Gelling solution, 100 μL: 94 μL monomer solution, 2 μL TEMPO (0.5%), 2 μL TEMED (10%), 2 μL APS (10%). To be prepared just before use, *see* Subheading 3.3, **step** 7.

Digestion buffer: In order to yield a final amount of 100 mL digestion buffer with Prot-K use TRIS (0.6057 g), EDTA (0.0372 g), Triton X-100 (100%, 500 μL), guanidinium chloride (7.6424 g). Fill up to 96 mL with Milli Q water. Adjust the buffer to pH 8.0. Store as 4.8 mL aliquots in −20 °C without Prot-K. Just before use add 100 μL ddH_2O and 100 μL Prot-K to one aliquot. Final concentration of Prot-K, 16 units/mL.

2.2 Small Items

Coverslips 22 × 22 mm^2, 18 × 18 mm^2.

Coverslip tweezers.

Double-sided tape of 100 μm thickness for creation of the gelation chamber.

24-well plate, e.g., μ-plate 24 Well Black with a well diameter of 14 mm (e.g., ibidi 82426).

Plastic dishes, 35 mm diameter with polymer coverslip bottom (e.g., μ-dish 35 mm, ibidi 81158).

Sharp, flat cutting tool, e.g., razor blade.

Soft brush.

2.3 Instruments

Shaker with temperature control.

LSFM, e.g., Ultramicroscope Blaze from Miltenyi Biotec or a custom-built setup featuring low and high numerical aperture (NA) water immersion (WI) objective lenses.

Long distance objectives with high NA, e.g., 25×, NA 0.95, working distance (WD) 8 mm; 12×, NA 0.53, WD 8.5 mm; 4×, NA 0.35, WD 16 mm.

Software for image processing, e.g., Fiji [13] or Imaris (Version 9.7.2 Bitplane Inc., Zurich, Switzerland).

3 Methods

We start out with mouse brain sections of various thicknesses that were fixed in 4% paraformaldehyde, and after fixation, they were transferred into 0.01% sodium azide and stored at +4 °C.

3.1 Immunostaining

1. Permeabilize the sample before immunostaining or expansion to improve penetration of the antibodies (*see* **Note 1**). To this end, incubate samples of thickness 0.2, 1, or 2 mm for 2, 12, or 36 h at 37 °C on a shaker with permeabilization buffer (*see* **Note 2**).

2. Wash three times with PBS at room temperature (RT) and incubate during each step for samples of thickness 0.2, 1, or 2 mm for 30, 45, or 60 min, respectively.
3. Block sample of thickness 0.2, 1, or 2 mm with blocking buffer (*see* **Note 3**) on a shaker at RT for 12, 24, or 48 h (*see* **Note 4**), respectively. Always use enough solution to cover the sample completely.
4. Incubate the sample (*see* **Note 5**) with the primary antibody (Ab) in blocking buffer on a shaker at 4 °C. Samples of thickness 0.2, 1, or 2 mm are incubated for 12, 24 h, or 4–5 days, respectively.
5. Wash the sample three times on a shaker at RT with blocking buffer. Samples of thickness 0.2, 1, or 2 mm are incubated for 30, 45, or 60 min.
6. Incubate the sample with the secondary Ab (*see* **Note 6**) in blocking buffer on a shaker at 4 °C. Incubate samples of thickness 0.2, 1, or 2 mm for 12 h, 1 day, or 4–5 days, respectively.
7. Wash the sample three times with PBS on shaker at room temperature (RT) and incubate during each step for samples of thickness 0.2, 1, or 2 mm for 30, 45, or 60 min, respectively.

3.2 Pre-expansion Imaging

1. Stain the cell nuclei (*see* **Note 7**) of the sample using Hoechst 33342 at a concentration of 2.5 μg/mL by incubating samples of size 0.2, 1, or 2 mm for 12 h, 2 days, or 3 days, respectively.
2. Wash the sample three times with PBS on shaker at room temperature (RT). Samples of size 0.2, 1, or 2 mm are incubated in each step for 30, 45, or 60 min, respectively.
3. For imaging, place the sample into a 24-well plate (e.g., 24 Well Black from ibidi 82426) or a μ-dish 35 mm filled with PBS depending on the size of the sample.

3.3 Sample Expansion

1. Dilute MA-NHS 1:500 in PBS from the 1 M aliquots, which are stored at −20 °C.
2. For linking, incubate the sample with 2 mM MA-NHS on a shaker at RT (*see* **Note 8**). Each sample is incubated in a separate well. Incubate samples of thickness 0.2, 1, or 2 mm with the linking solution for 2 h, 1 day, or 2 days, respectively.
3. Wash the sample three times with PBS on a shaker at room temperature (RT). Incubation times in each washing step for samples of size 0.2, 1, or 2 mm are 30, 45, or 60 min, respectively.
4. During the last washing step, thaw the monomer solution aliquot (stored at −20 °C) and put it on ice.

5. Incubate with monomer solution on a shaker at 4 °C (*see* **Note 9**). Samples of size 0.2, 1, or 2 mm are incubated for 2 h, 1 day, or 2 days, respectively.
6. Prepare a "gelation chamber" for the sample, while it is incubated with the monomer solution. To this end, use a coverslip (22 × 22 mm^2, *see* **Note 10**) as a base and place spacers on two opposite corners (*see* Fig. 1). The spacers could be made with a double-sided tape, which have a thickness of about 100 μm and 9 mm width by folding it as many times as needed until the desired thickness is obtained. Then cut it into a rectangular shape with a size of about 6 × 9 mm^2.
7. Mix the monomer solution with TEMPO and TEMED, and add ammonium persulfate (APS) at last (*see* **Note 11**), and vortex briefly to prepare the gelling solution. For 100 μL of gelling solution, use 94 μL monomer solution, 2 μL TEMPO (0.5%), 2 μL TEMED (10%), and 2 μL APS (10%).
8. Remove the monomer solution from the sample and apply the gelling solution. Incubate the sample with the gelling solution for 5–10 min on a shaker at 4 °C.
9. During the incubation, remove the cover tape from the spacers.
10. Add a few drops of gelling solution into the middle of the bottom of the prebuilt gelling chamber with the spacers and keep the chamber at 4 °C. Then transfer mouse brain slices using a brush (*see* **Note 12**). Make sure the sample rests on the bottom, and add gelling solution around the sample (*see* Fig. 2 and **Note 13**).
11. Cover the sample with a smaller (e.g., 18 × 18 mm^2) coverslip (*see* **Note 14**). Avoid the formation of bubbles around or near the specimen. There should be enough gelling solution around the sample so that it can be removed later and the sample is still surrounded by the gel (*see* Fig. 3).
12. Place the gelation chamber on a small weighing boat (*see* **Note 15**) and place it in a moist chamber to prevent the gel from drying out. To improve penetration of the gelling solution into large samples before polymerization, it is advisable to keep the samples at 4 °C for several hours. For samples of 2 mm thickness, an incubation for 12 h is recommended.
13. Incubate for 2 h at 37 °C. Make sure that the samples do not dry out.
14. Remove the coverslip and excess gel: To remove the upper coverslip from the spacer, use a wet brush with PBS-T and apply on the spacers (*see* **Note 16** and Fig. 4). After a couple of seconds, remove the coverslip and the spacers with forceps. Remove the excess of the gel around the tissue with a scalpel or razor blade (*see* **Note 17**).

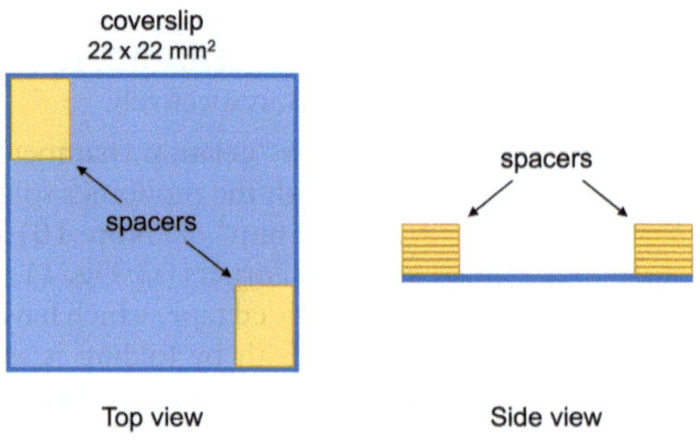

Fig. 1 Placement of the spacers on the coverslip

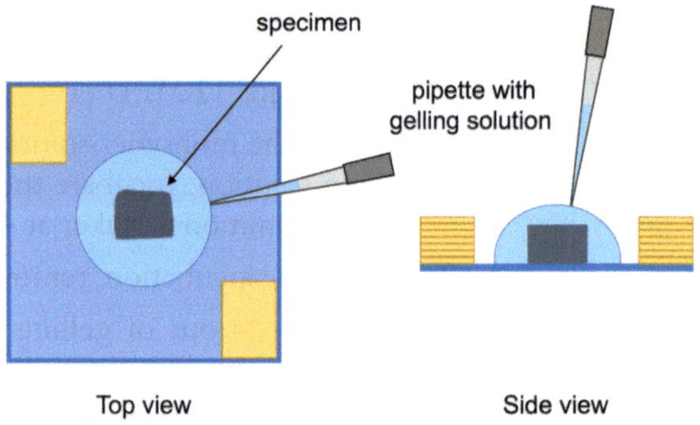

Fig. 2 Placement of the sample in the gelation chamber

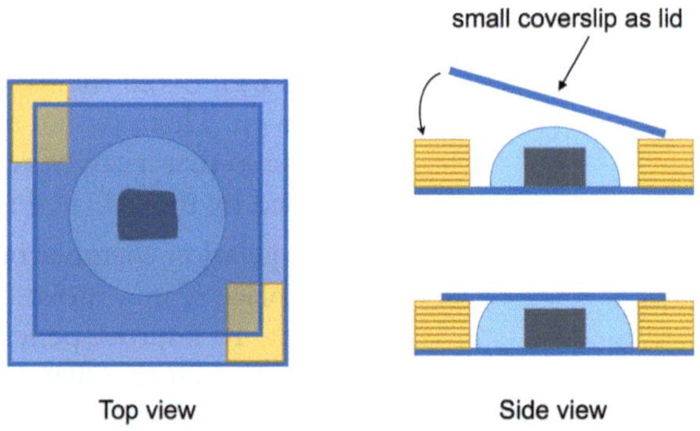

Fig. 3 Closure of the gelation chamber

15. Digest the sample homogeneously with proteinase K. Fill little plastic bowls, e.g., little weighing boats, with digestion buffer at 37 °C and add the gel-embedded sample (*see* **Note 18**). Use enough digestion buffer in order to cover the gel block completely (about 2.5 mL for each weighing boat). Place into a wet chamber and keep the sample in digestion buffer until it is completely transparent. The exact duration depends

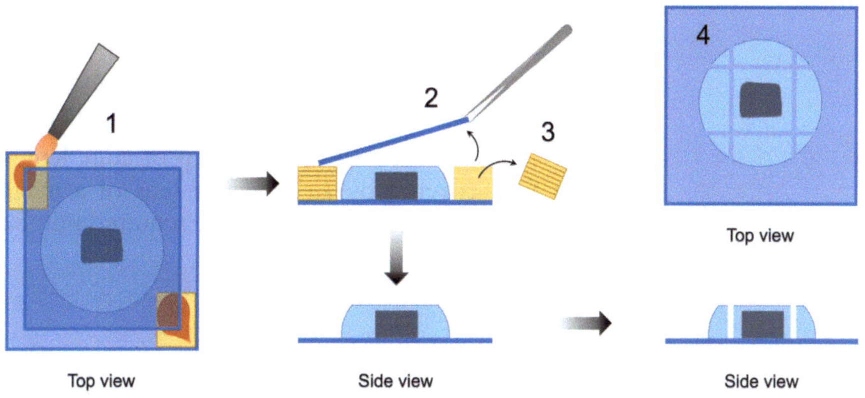

Fig. 4 Preparation of the specimen for digestion. (1) Use a wet brush with PBS-T to remove the upper coverslip from the spacer, (2) remove the top coverslip and (3) the spacer with forceps. (4) Finally, remove the excess gel near the gel-embedded specimen using a sharp cutting device

on the specific tissue. It takes about 24 h for a 200-μm-thick mouse brain slice and up to 4 days for a 2-mm-thick mouse brain slice. Change the digestion buffer every 24 h if a long incubation duration is required.

16. Carefully remove the digestion buffer with a pipette. Use the bottom coverslip to transfer the sample into a 6-well plate (*see* **Note 19**), and depending on the desired expansion factor, wash the samples thoroughly: for 1.5× expansion, wash with PBS four times, each time for 30 min; for 4× expansion, wash with 5 mM Hepes, pH 7.4. For a 0.2-mm-thick, 1-mm-thick, or 2-mm-thick mouse brain slice, wash four times for 30, 45, or 60 min, respectively. Then wash a final time (*see* **Note 20**).
17. To stain the cell nuclei, add Hoechst 33342 (H3570, Invitrogen) at a final concentration of 2.5 μg/mL and incubate for samples of thickness 0.2, 1, or 2 mm for 12 h, 2 days, or 3 days, respectively (*see* **Note 21**). Then, transfer the sample to the imaging chamber for microscopy.

3.4 Imaging from the Meso- to the Nanoscale

Generally, the physical sample extension exceeds the field of view of the microscope. In that case, the sample must be imaged in mosaic tiles and subsequently stitched together to yield the complete image. To enable the stitching, neighboring tiles must have a 10% overlap with each other. We characterize the sample size by its length, l, width, w, and height, h.

Mesoscopic imaging serves to obtain an overview of the sample. From the obtained images, the regions, which need to be analyzed in greater detail, can be selected.

1. For sample mounting, coat the bottom of the imaging chamber with 0.1% poly-l-lysine for 5 min to avoid movements of the sample during the measurement.

2. Prepare the sample for imaging: slowly let the cleared and expanded gel piece with the specimen slide into the coated imaging chamber. Apply the respective expansion solution to the imaging chamber to prevent drying of the gel.
3. Using a detection camera with a chip of edge length, l_c, the object field covered by an objective lens of a magnification M corresponds to $\frac{l_c^2}{M^2}$. Determine the required number of tiles using the formula:

$$n_{\text{tile}} = 1.44 \frac{M^2 \cdot l \cdot w}{l_c^2}$$

Here, we considered a 10% overlap at each tile border for successful tile stitching.
4. Acquire the respective number of image tiles with a low-magnification objective (e.g., a 4× water WI objective lens, NA 0.35) of the sample (*see* **Note 22**).
5. Determine the required number of planes n_{steps} in order to image your sample in accordance with the Nyquist criterion in axial direction and again choosing 10% overlap according to [14].

$$n_{\text{steps}} = 0.9 \frac{h \cdot \text{NA}^2}{\lambda}$$

(*see* **Note 23**). Here, λ designates the wavelength of fluorescence emission maximum.
6. To acquire high resolution images from the expanded sample, choose a high-NA objective lens for imaging, e.g., a 25× objective with a of NA 0.95 (*see* **Note 24**). The given formulas show that a sample of 1 cubic millimeter (after expansion, *see* **Note 25**) now requires 9 tiles in each plane (actually only 5.16, what is not feasible) and nearly 1500 axial sections. This corresponds to about 113 GB of data.
7. For representation of the complete data, the intensity histograms were first adjusted to normalize brightness and contrast throughout the complete dataset.
8. Stitch the single image stacks using the Imaris Stitcher (Oxford Instruments) or by using the stitching plugin of Fiji [15] (*see* **Note 26**).
9. Finally, possible intensity variations of the sample along the axial direction may be compensated to improve the contrast throughout the 3D stack. For this purpose, a histogram normalization should be performed in every image plane.

4 Conclusions

The described workflow provides a standardized and reproducible method for the expansion and imaging of neuronal tissue, enabling detailed mesoscopic mapping of neuronal connections with the option of highlighting details of interest in super-resolution. We used the described methods very successfully in the past years for a variety of samples, such as neuronal organoids (*see* Figs. 5 and 6), midbrain–striatum assembloids, extended mouse brain sections, and human glioblastoma tumors. Our workflow will allow researchers to gain a deeper understanding of the complex circuitry underlying brain function and pathology, ultimately leading to improved diagnosis and treatment of neurological disorders.

5 Notes

1. We present here the steps for labeling by immunofluorescence only. For use of, e.g., SNAP-, HaloTags, or only autofluorescent proteins, the protocol has to be accordingly modified.
2. For thicker or more dense samples, it is recommended to increase the CHAPS concentration, e.g., 1% w/v for 2-mm-thick mouse brain slices.
3. For better results, use serum corresponding to the host species of the secondary Ab you will use.
4. For longer blocking sessions, place the sample on a shaker at 4 °C.
5. After removing the blocking solution, it is not necessary to wash the samples.
6. When using more than one primary and corresponding secondary antibody, we recommend immunolabeling in a serial manner instead of a parallel combination of antibodies to avoid cross-reactivity between antibodies.
7. We usually perform a nuclear staining to enable a pre-expansion imaging to obtain an overview of the sample. Alternatively, this staining could be done after sample digestion.
8. For linking, remove the PBS carefully from the well, or transfer the sample to another well with linking solution using a pipette coated with PBS-T to prevent the samples from sticking. Use a coated pipette tip to transfer small and delicate samples, e.g., organoids, or a brush for bigger samples. When pipetting, it is recommended that the pipette tip be trimmed to have a larger diameter.

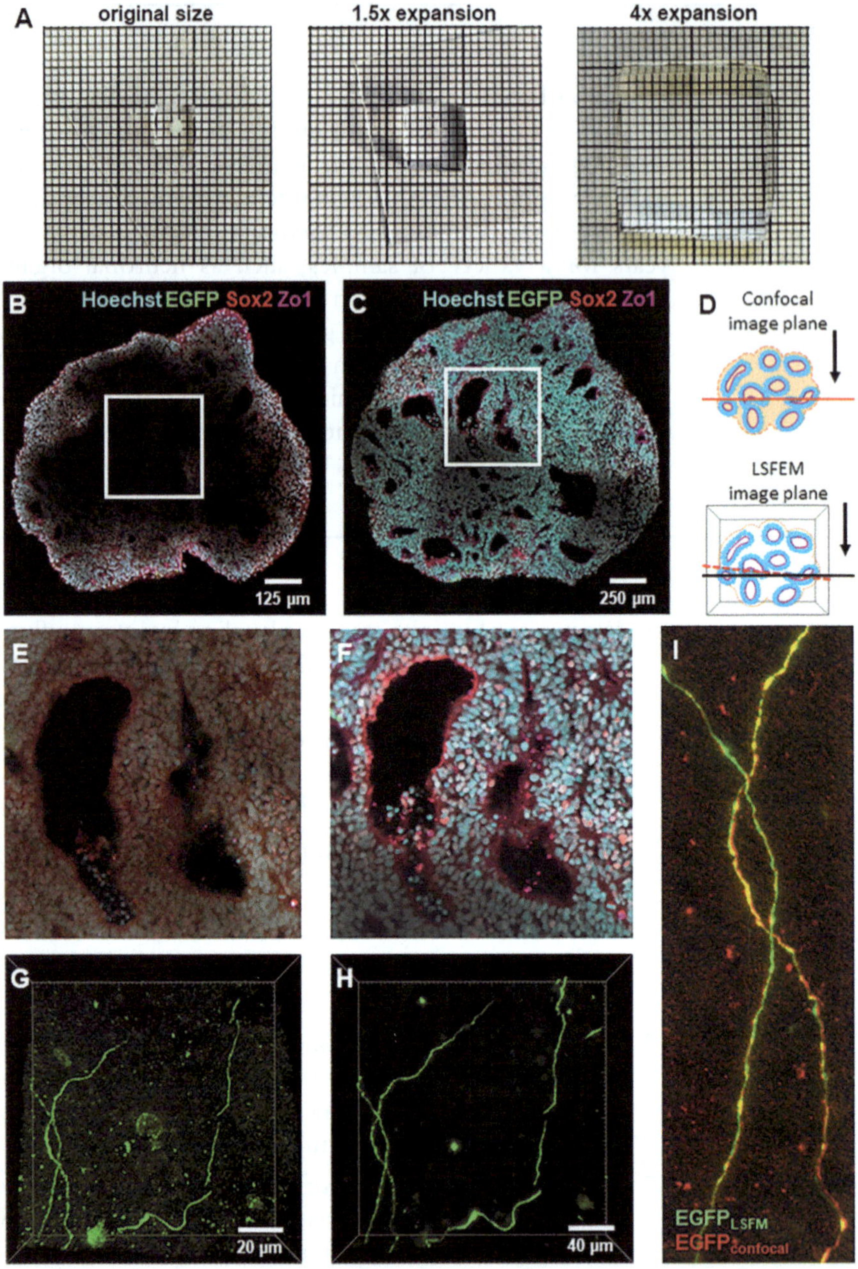

Fig. 5 Analysis of organoid sample preparation and structural integrity post LSFEM processing. (**a**) Visualization of the preparation process of a 2-month-old brain organoid step by step: starting with embedding in polyacrylamide gel, followed by proteinase K digestion leading to organoid clearing with approximately 1.5-fold expansion, and concluding with bi-distilled water treatment resulting in about four-fold expansion. (**b–i**) Analysis of the clearing procedure's effect on both staining fidelity and structural integrity within a 40-day-old brain organoid containing 10% EGFP-positive cells. The organoid was stained with Hoechst (cyan) for cell nuclei, Sox2 (red) for progenitor cells, and ZO1 (magenta) for tight junctions, imaged before and after the clearing process (i.e., with 1.5-fold expansion). (**b**) Optical section of the organoid before clearing, captured at 1.2 mm depth using a Zeiss LSM 880 microscope (25× NA 0.8 water immersion objective). (**c**) Optical slice after clearing showing 1.5-fold expansion, imaged with LSFM (10× NA 0.3 WI objective). (**d**) Image fields in (**b**) and (**c**) are not identical yet contain a shared region (**e**, **f**) attributable to a minor sample rotation during

9. It is important to have the samples on a shaker to improve penetration and thus the linking of the polyacrylamide matrix within the sample. If a cooled shaker is not available, shake manually approximately every 15–20 min.
10. The size is not mandatory. The coverslip just needs to be large enough to hold the sample even after expansion.
11. The gelling solution needs to be prepared just before being applied, and all the solutions have to be kept on ice.
12. To facilitate the transfer of samples to the coverslip, it is recommended to first pipette one or two drops of the gelling solution onto it. This helps to spread the tissues when transferring the sample. Use a coated pipette tip to transfer small and delicate samples, e.g., organoids, or a brush for bigger samples such as mouse brain sections. When pipetting, it is recommended that the pipette tip should be trimmed to have a larger diameter. It is necessary to remove the excess solution until the sample touches the bottom of the coverslip, and then add gelling solution to cover the sample.
13. To keep thin sections (up to 100 μm) as flat as possible, samples can be transferred to a coverslip coated with poly-l-lysine to keep it attached to the bottom. The total volume in μL of gelling solution needed to cover the sample can be estimated as follows:

$$V\,[\mu\mathrm{L}] = (\mathrm{spacer} - \mathrm{thickness}\,[\mathrm{mm}] \times \mathrm{area\ top\ coverslip}\,[\mathrm{mm}])/2000.$$

To avoid polymerization of the sample inside the well, place the well plate with the samples on ice during the process. In addition, the simultaneous preparation of many samples should be avoided. If the transfer to the gelling chamber is slow, there is a risk that the gel polymerizes in the well plate.

14. The upper coverslip used as a lid should be smaller, as this facilitates removal after gelation. When the spacers are very thick, it may happen that the solution does not touch the upper coverslip. This can be solved by adding more gelling

Fig. 5 (continued) manipulation and transfer to the LSFM chamber. This rotation may cause a slight tilt in the LSFM-imaged optical section relative to that obtained with the LSM880. (**e**, **f**) Detailed views of the region common to (**b**) and (**c**), respectively. Image (**e**) originating from (**b**) was contrast enhanced for improved visibility. (**g**) 3D rendering of the EGFP-channel from a volumetric stack (212 × 212 × 39 μm^3) captured with the Zeiss LSM 880 (40× NA 1.1 WI objective) from the same sample. (**h**) Post-digestion and 1.5-fold expansion 3D rendering of the same volume imaged using LSFM (25× NA 1.1 WI objective), demonstrating not only the preservation of EGFP fluorescence but also superior data quality post-expansion. (**i**) Direct overlay of two neurites for comparison from (**g**) and (**h**) confirming the structural correspondence between the pre-expansion and post-1.5-fold expanded samples. (Adapted from Rodriguez-Gatica et al. [12])

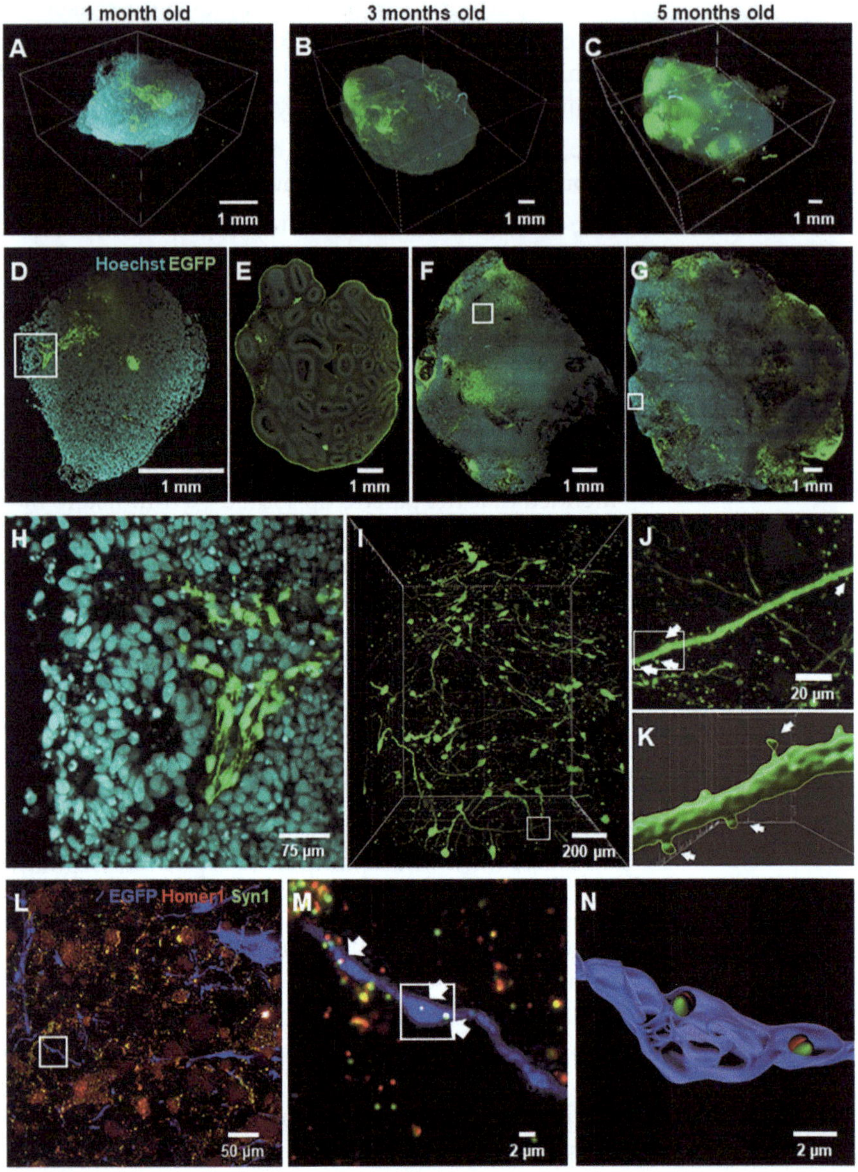

Fig. 6 Comprehensive analysis of chimeric brain organoids with 10% EGFP-expressing cells over time. (**a**–**c**) Three-dimensional representations of brain organoids at various stages: 1-month-old (**a**), 3 months (**b**), and 5 months (**c**) with imaged volumes of 3.6 × 3.6 × 1.6 mm^3, 8.6 × 11.7 × 4.4 mm^3, and 13.1 × 14.9 × 5.2 mm^3, respectively. (**d**–**g**) LSFM images of cleared and 1.5-fold expanded organoids showcasing 10% EGFP-expressing cells (green) and nuclei stained with Hoechst (cyan) at different time points: 1 month (**d**), 3 months (**e**), 5 months (**f**), and 14 months (**g**). Optical sections were captured at depths of 880, 1140, 1830, and 1120 μm with corresponding image sizes of 3.6 × 3.6 mm^2, 8.6 × 11.7 mm^2, 13.1 × 14.9 mm^2, and 16.6 × 17.3 mm^2, respectively. These data enable a comparative analysis of the rosette structures' morphology and the distribution of GFP-positive cells throughout the entire volumes over time. (**h**) Magnification of the region highlighted in (**d**), measuring 520 × 520 μm^2, illustrating cell distribution and the preservation of autofluorescent protein fluorescence providing insights into cellular organization. (**i**) Rendering of a 3D stack from the area marked in (**f**) with dimensions of 1248 × 1548 × 1275 μm^3 imaged after fourfold expansion using a 25× NA1.1 WI objective lens, further emphasizing structural details and protein preservation. (**j**) Maximum intensity projection (MIP) of the area indicated in (**i**), 185 × 132 μm^2, displaying spine-like

solution from the side using a small (e.g., 200 μL) pipette. The upper coverslip (i) creates a planar surface of the gel and (ii) stabilizes the gelling solution over the specimen.

15. This is to prevent the gel from coming into direct contact with water, as it will absorb it and could begin to expand, damaging the sample.
16. This serves to moisten the double-sided tape and make it easier to remove.
17. Do not remove the sample from the bottom coverslip! This might damage the gel and sample. The cutting should be done perpendicular to the bottom coverslip in order to create a clean interface. This improves the illumination conditions for light sheet illumination from the side, because there is a small deviation of the refractive indices of the gel and surrounding buffer or water.
18. For thicker samples, the gelated sample can be placed into a 6-well plate.
19. It is important to keep a coverslip under the sample (24 × 24 mm for a 6-well plate), in case it is necessary to transfer it again. The slice thicknesses given refer to the slice thickness before expansion.
20. For 4× expansion, the washing time depends on the size of the gel. Thicker gels take longer to expand, so keep them until they reach the desired size.
21. The expansion process leads to a reduction of the previous Hoechst staining density; this is why the sample should again be stained here.
22. Assuming a standard sCMOS camera with 2048 pixels of size 6.45 μm per line, the object field covered corresponds to an area of 3.3 × 3.3 mm^2. Please note that this data collection does not meet the Nyquist criterion.
23. We assumed here that the sample is imaged in water or buffer with a refractive index of 1.33. For further details, you may also check https://svi.nl/NyquistCalculator and https://svi.nl/NyquistRate#Equations.

Fig. 6 (continued) structures (indicated by arrows). (**k**) Surface rendering highlights neural projections with spine-like structures offering a closer look at the intricate cellular morphology. (**l**) MIP of 20 layers (approximately 300 μm in axial depth) presenting EGFP in blue, and synaptic proteins SYN1 and HOMER1 in green and red, respectively. (**m**) Magnification of the region marked in (L) displaying the colocalization of pre- and post-synaptic proteins SYN1 and HOMER1 at axonal boutons (arrows point to synapses along neurites) demonstrating the existence of synaptic junctions. (**n**) 3D reconstruction of the multi-synaptic bouton featured in the boxed region of (**m**) providing a detailed visualization of synaptic architecture. (Adapted from Rodriguez-Gatica et al. [12])

24. The choice of the correct lens is of utmost importance for high-resolution imaging. Clearly, a high NA is required; however, for imaging expanded samples, a long working distance (WD) is also essential, because the expanded samples are large. Unfortunately, the selection of commercially available lenses with high NA and long WD is quite small. In any case, such objectives are quite voluminous.
25. A sample of 1 cubic millimeter before expansion corresponds after full expansion to a volume of $4 \times 4 \times 4 = 64$ cubic millimeters. Now, 81 tiles are required to image one plane, and 5900 in axial direction. This corresponds to 4 TB of data.
26. To simplify stitching, especially when datasets exceeded the available computer RAM, the process was performed in two steps. First, substacks of the 3D datasets were created using a Fiji script. Each substack contained about 15% of the images in the center of the full stack. Each substack was stitched to its respective neighboring substack, yielding the best overlap in terms of the cross-correlation measure. Based on the localization information of each substack after stitching, the full 3D stacks were stitched.

Conflict of Interest

The authors have no conflicts of interest to declare that are relevant to the content of this chapter.

References

1. Schwarz MK, Remy S (2019) Rabies virus-mediated connectivity tracing from single neurons. J Neurosci Methods 325:108365
2. Mano T, Albanese A, Dodt H-U et al (2018) Whole-brain analysis of cells and circuits by tissue clearing and light-sheet microscopy. J Neurosci 38:9330–9337
3. Chen F, Tillberg PW, Boyden ES (2015) Optical imaging. Expansion microscopy. Science 347:543–548
4. Lillvis JL, Otsuna H, Ding X et al (2022) Rapid reconstruction of neural circuits using tissue expansion and light sheet microscopy. Elife 11:e81248
5. Bürgers J, Pavlova I, Rodriguez-Gatica JE et al (2019) Light-sheet fluorescence expansion microscopy: fast mapping of neural circuits at super resolution. Neurophotonics 6:015005
6. Gao R, Asano SM, Upadhyayula S et al (2019) Cortical column and whole-brain imaging with molecular contrast and nanoscale resolution. Science 363:eaau8302
7. Düring DN, Rocha MD, Dittrich F et al (2019) Expansion light sheet microscopy resolves subcellular structures in large portions of the songbird brain. Front Neuroanat 13:2
8. Stockhausen A, Rodriguez-Gatica J, Schweihoff J et al (2023) Airy beam light sheet microscopy boosted by deep learning deconvolution. Opt Express 31:10918–10935
9. Stockhausen A, Bürgers J, Rodriguez-Gatica JE et al (2020) Hard-wired lattice light-sheet microscopy for imaging of expanded samples. Opt Express 28:15587–15600
10. Feng R, Wang D, Chen Y et al (2021) Protocol for constructing a versatile tiling light sheet

microscope for imaging cleared tissues. Star Protoc 2:100546
11. Truckenbrodt S (2023) Expansion microscopy: super-resolution imaging with hydrogels. Anal Chem 95:3–32
12. Rodriguez-Gatica JE, Iefremova V, Sokhranyaeva L et al (2022) Imaging three-dimensional brain organoid architecture from meso- to nanoscale across development. Development 149:dev200439
13. Schindelin J, Arganda-Carreras I, Frise E et al (2012) Fiji: an open-source platform for biological-image analysis. Nat Methods 9:676–682
14. Kubitscheck U (2017) Principles of light microscopy. In: Kubitscheck U (ed) Fluorescence microscopy. Wiley-VCH, Weinheim, pp 23–83
15. Preibisch S, Saalfeld S, Tomancak P (2009) Globally optimal stitching of tiled 3D microscopic image acquisitions. Bioinformatics 25:1463–1465

Chapter 5

Expansion Lattice Light-Sheet Microscopy Reveals Subcellular Organization and Molecular Profile of the Brain with Nanoscopic Resolution

Wei Wang and Ruixuan Gao

Abstract

Light-sheet fluorescence microscopy (LSFM) uses a sheet of light to spatially confine excitation of fluorophores to a narrow cross-section of the sample. Lattice light-sheet microscopy (LLSM) is an emerging LSFM method that takes advantage of light-sheet generated by an optical lattice. The ultra-thin illumination of LLSM serves to enhance the axial resolution and reduce the photobleaching effect of conventional LSFM methods. When combined with tissue clearing and expansion, LLSM allows fast volumetric imaging of cellular and subcellular morphologies across a large sample size with nanoscale resolution and molecular contrast. In this chapter, we introduce a recently developed method named expansion lattice light-sheet microscopy (ExLLSM or ExL) to demonstrate the advantage of this combination. Using ExL, we show the molecular imaging of subcellular morphologies across the mouse cortex and *Drosophila* brain with sub-diffraction-limit resolution. We further describe the technical details of ExL to enhance its accessibility for the imaging and neuroscience communities.

Key words Light-sheet microscopy, Lattice light-sheet microscopy, Expansion microscopy, Tissue expansion, Tissue clearing, Super-resolution microscopy

1 Introduction: Expansion Lattice Light-Sheet Microscopy

With bright and molecularly specific fluorescent tags, fluorescence microscopy allows direct and multiplexed observation of the molecular profile of biological structures and processes. However, owing to the diffraction limit of light, subcellular features of ~300 nm or smaller in size, such as presynaptic terminals, postsynaptic dendritic spines, and intracellular organelles, can hardly be imaged at their original length scale using conventional fluorescence microscopy.

To overcome this diffraction limit, super-resolution imaging techniques, such as PALM (photo-activated localization microscopy) [1], STORM (stochastic optical reconstruction microscopy) [2, 3], STED (stimulated emission depletion) microscopy [4–6], RESOLFT (reversible saturable optical fluorescence transitions)

Ludovico Silvestri (ed.), *Light Sheet Microscopy*, Neuromethods, vol. 229, https://doi.org/10.1007/978-1-0716-5100-1_5,

[7, 8], SIM (structured illumination microscopy) [9, 10], and MINFLUX (minimal photon fluxes) [11–13], have been developed to capture nanoscale features in a broad range of biological specimens. However, such super-resolution methods often face challenges when imaging thick tissue samples. Specifically, conventional super-resolution imaging methods are often limited in the imaging depth due to the strong scattering effect of the tissue and the short working distance of high-numerical-aperture objectives. In addition, in some super-resolution imaging methods, a high illumination power is needed to quench or switch the fluorophores, which can introduce additional photobleaching effect to the sample. Moreover, the acquisition time of each imaging frame can take up to tens of minutes, making it challenging to scale up to a large tissue volume.

Given these challenges, an imaging method that readily captures the subcellular morphology and molecular profile across a large tissue volume has been sought after in the neuroscience as well as other biomedical fields. In this chapter, we introduce a recently developed imaging method named expansion lattice light-sheet microscopy (ExLLSM, or ExL for short) to address this methodological gap. By combining lattice light-sheet microscopy (LLSM) and expansion microscopy (ExM), we show that ExL enables 3D mapping of proteins across the thickness of the mouse cortex and the entirety of the *Drosophila* brain with nanoscopic resolution (Fig. 1).

1.1 Lattice Light-Sheet Microscopy

Light-sheet fluorescence microscopy (LSFM) is a powerful tool for fast and scalable volume fluorescence imaging. In LSFM, a thin sheet of light is generated to illuminate a planar cross-section of the sample [14–16]. Since this sheet of light only excites fluorophores within its illumination plane, LSFM prevents excessive excitation from out-of-plane illumination. Hence, using LSFM, one can perform large-scale volume imaging of biological samples with reduced photobleaching and phototoxicity. In addition, LSFM setups are often equipped with a fast 2D image detector, such as a scientific complementary metal-oxide-semiconductor (sCMOS) camera, to capture the entire illuminated plane. Along with its planar illumination configuration, this detection setup allows LSFM to sequentially sweep through a large volume with enhanced speed [15]. In the context of 3D volume imaging of expanded and optically cleared tissue samples, the low photobleaching effect and the high imaging throughput make LSFM an ideal platform to perform the imaging of such samples.

In practice, a popular option for LSFM illumination is a Gaussian light-sheet, which can be generated using a cylindrical lens. One limitation of Gaussian light-sheet illumination is, however, its difficulty in maintaining high axial resolution over a large field of view. This issue can be solved by, for example, switching the illumination

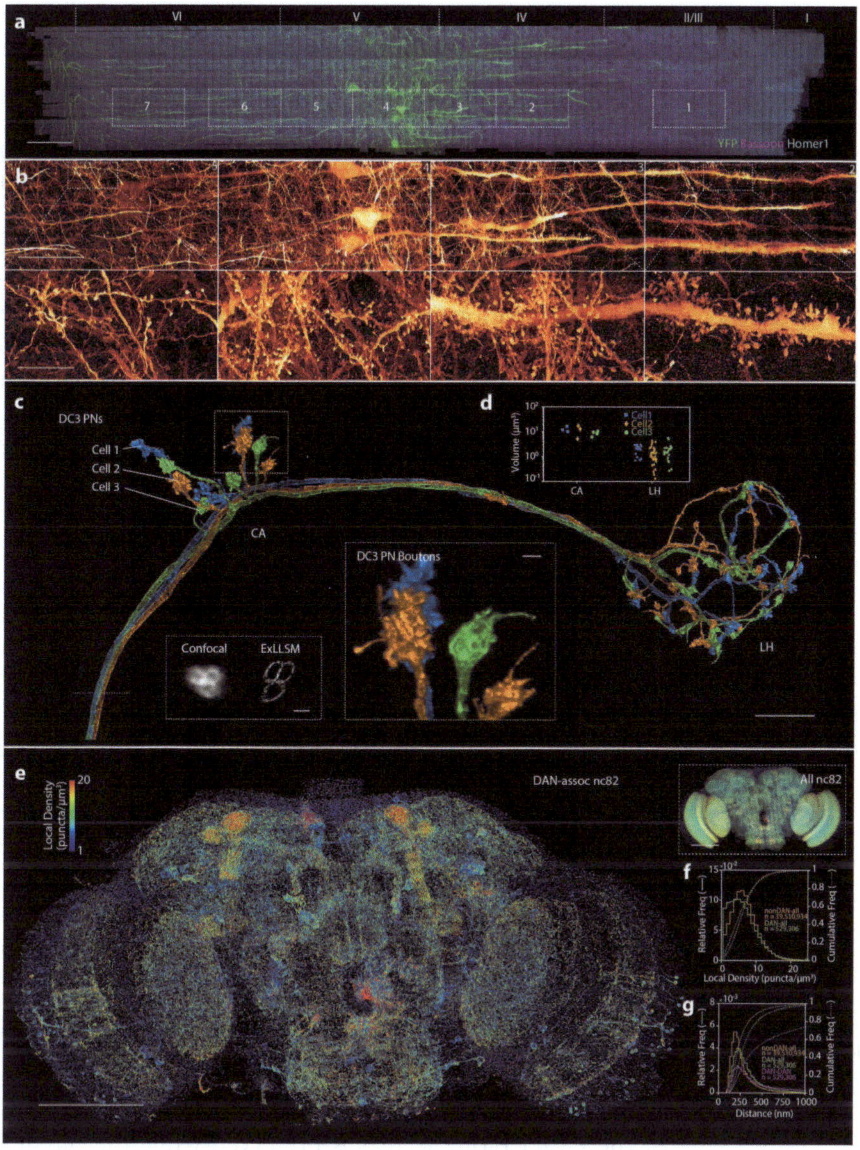

Fig. 1 Biological data obtained by expansion lattice light-sheet microscopy (ExLLSM or ExL). (**a**) Maximum intensity projection (MIP) view of a Thy1-YFP (green) mouse tissue slice (~1900 × 280 × 70 μm) spanning from the pia to the white matter of the primary somatosensory cortex, immunostained against synaptic proteins Bassoon (magenta) and Homer1 (cyan). Scale bar, 100 μm. Unless otherwise noted, here and after, all the scale bars are in the pre-expansion dimension. (**b**) (Top) MIP views showing YFP expressing neurons (orange) from the white boxes (2, 3, 4, 5) in (**a**) and (bottom) magnified views of the boxed regions in the top panels showing different spine morphologies. Scale bars, (top) 50 μm and (bottom) 10 μm. (**c**) MIP view showing three olfactory projection neurons (PNs) originating at the DC3 glomerulus of the antennal lobes of an adult *Drosophila* brain and partially traced to the calyx (CA) and lateral horn (LH). Scale bar, 10 μm. (Inset, white box) Cross-section of the axon bundle imaged by (left) confocal microscopy and (right) ExL. Scale bar, 1 μm. (Inset, yellow boxes) A magnified view of DC3 PN boutons in CA. Scale bar, 1 μm. (**d**) Plot showing the volume of individual DC3 PN boutons in CA and LH. (**e**) MIP view of a subset of nc82 puncta associated with dopaminergic neurons (DAN-associated nc82) in an adult *Drosophila* brain. The color indicates the local

pattern to propagation-invariant beams, such as a Bessel beam, Airy beam, or an optical lattice of these beams. When LSFM uses a light-sheet illumination constructed from an optical lattice of such beams, it is called lattice light-sheet microscopy (LLSM) (Fig. 2) [17]. To highlight the differences between lattice and non-lattice illumination, cellular dynamics of filopodia have been captured with the same excitation power using a hexagonal optical lattice and a Bessel beam (in the SIM mode). While the two beam profiles have achieved comparable resolution, all the cells imaged using the lattice illumination have shown minimal bleaching, indicating that much lower laser intensity is needed to achieve a similar signal-to-noise ratio. Furthermore, when a dithered mode of the lattice illumination (i.e., lattice light-sheet illumination) and a swept Bessel beam light-sheet illumination are applied to the same sample, the lattice light-sheet illumination has demonstrated multiple advantages over its non-lattice counterpart. First, the dithered lattice light-sheet illumination has shown a lower imaging background, suggesting the excitation is confined to a narrower region. Second, the dithered lattice light-sheet illumination has achieved a higher axial resolution compared to the swept Bessel beam light-sheet illumination.

In summary, lattice illumination shows a number of advantages over conventional illumination for high-resolution, low-photobleaching imaging [18]. Using lattice light-sheet illumination can improve axial resolution while reducing the photobleaching effect. Given these advantages, LLSM has been applied to image and reconstruct (sub)cellular structures and processes of cell cultures, tissues, embryos, and organisms in their native states, often at sub-second intervals, for hundreds to thousands of time points [17, 19–21].

1.2 Expansion Microscopy

Expansion microscopy (ExM) breaks the diffraction limit of light via physical swelling of polymerized biological samples (Fig. 3) [22, 23]. During this sample expansion, relative spatial information of the target biomolecules is preserved while the physical dimensions of the sample are increased multiple folds. For example, if the imaging resolution prior to the expansion is ~300 nm, the post-expansion effective resolution becomes ~75 nm after an

Fig. 1 (continued) puncta density. Scale bar, 100 μm. (Inset) MIP view of all nc82 puncta using the same color coding of local density. Scale bar, 100 μm. (**f**) Plot showing the distribution of local densities of DAN-associated nc82 puncta (green) and non-DAN-associated nc82 puncta (orange) in (**e**). (**g**) Plot showing distribution of distances from DAN-associated nc82 puncta (green) and non-DAN-associated nc82 puncta (orange) to the nearest nc82 punctum of any kind, and the nearest neighbor distances from one DAN-associated nc82 to another (magenta). (Reproduced/modified from Gao et al., *Science*, https://doi.org/10.1126/science.aau8302 (2019), AAAS [43].)

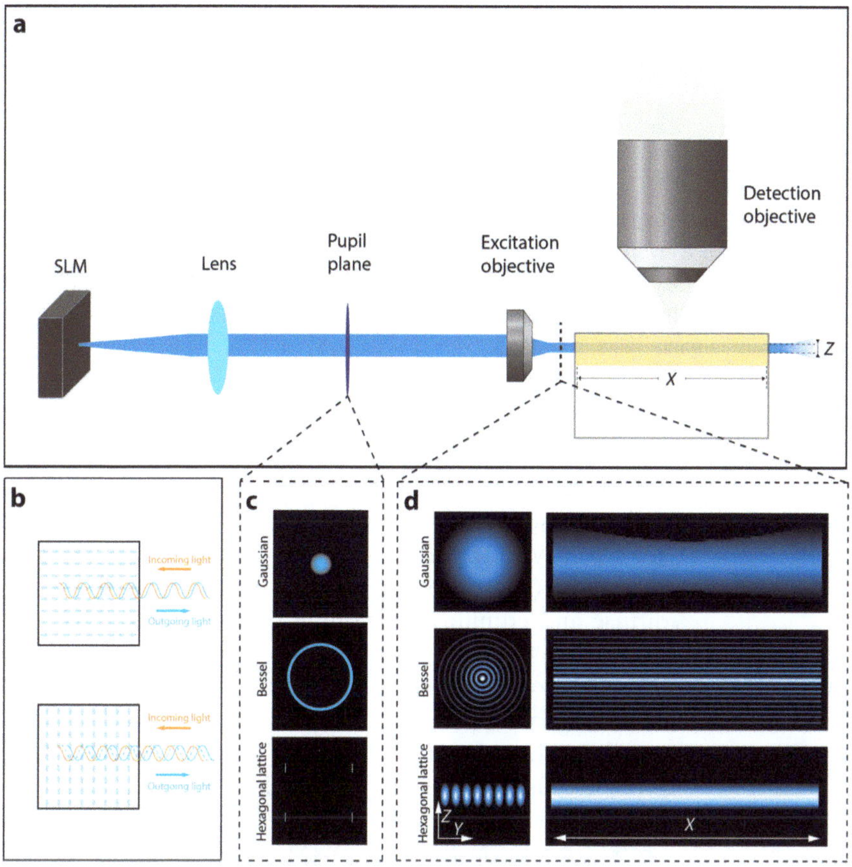

Fig. 2 Generation of lattice light-sheet for ExL. (**a**) A simplified schematic showing the basic optical setup of lattice light-sheet microscopy (LLSM). (**b**) Spatial light modulator (SLM) used in (**a**) is a liquid crystal device, in which electric signals can modulate the phase and intensity of the light beams passing through or reflecting on each pixel of the device by changing effective refractive index of the liquid crystal. The two images show how the SLM can affect the phase of incoming and outgoing light beams via interference when it is off (top) and on (bottom). (**c**) Intensity at the rear pupil from Gaussian beam (top), Bessel beam (middle), and hexagonal lattice (bottom) illumination. (**d**) Beam profile of the Gaussian beam (top), Bessel beam (middle), and hexagonal lattice (bottom) illumination in XZ and Y

approximately fourfold sample expansion. Unlike most super-resolution imaging methods, ExM does not require modification of the microscope. This makes ExM compatible with virtually any diffraction-limited fluorescence microscopes as well as super-resolution imaging techniques.

To prepare an ExM sample, a swellable hydrogel is first formed across the sample to create an expandable hydrogel–sample composite. We commonly use hydrogels formed via free-radical chain-growth polymerization, such as a co-polymer of acrylamide and sodium acrylate. After adding a free-radical initiator [e.g., ammonium persulfate (APS)] to the gelling solution, the hydrogel monomers (e.g., acrylamide and sodium acrylate) and crosslinkers (e.g.,

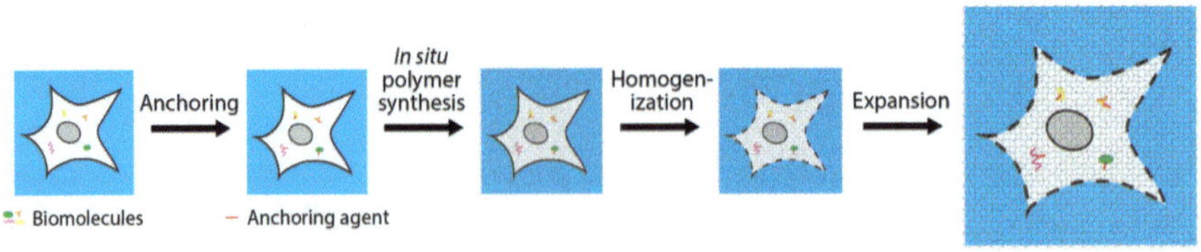

Fig. 3 General workflow of expansion microscopy (ExM). Biological samples are first incubated with an anchoring reagent before a hydrogel polymer network (with biomolecules of interest anchored) is formed across the sample in situ. Next, in the homogenization step, the molecular connectivity that has maintained the mechanical integrity of the biological sample is broken down. Finally, the sample–hydrogel composite is expanded isotropically multiple folds in water

bis-acrylamide) react with each other via free-radical chain-growth polymerization to form a highly crosslinked hydrogel network across the sample (with the target biomolecules anchored to the hydrogel, see later discussion). These monomers and crosslinkers are typically much smaller in size compared to the native macromolecules (e.g., proteins) within the sample and therefore can diffuse and equilibrate throughout the sample prior to the polymerization to form a uniform hydrogel network. By design, the ExM hydrogel shows a remarkable capability of swelling after absorbing pure water because of the entropy effect as well as the repelling charges on the polymer chains [24–27]. This swelling process causes the hydrogel–sample composite to increase several folds in size nearly isotropically. For example, the widely used protein-retention ExM (proExM) protocol physically enlarges tissue samples approximately four times in size and close to 100 times in volume [28, 29].

From an imaging perspective, a robust ExM protocol needs to ensure isotropic expansion as well as sufficient retention of the target biomolecules or labels. Isotropic sample expansion, however, cannot happen without additional sample treatments post-gelation because the fixed tissue sample often maintains its chemical and physical connectivity at the molecular level after polymerization. Expansion of such an untreated sample–hydrogel composite can result in macroscopic expansion artifacts such as sample stretching, deformation, and cracking, as well as microscale and nanoscale sample distortions (*see* [23] for a more in-depth discussion on the nanoscale distortion of ExM and expansion isotropy). Thus, a homogenization step (also called a digestion step) is needed to break down the aforementioned molecular connectivity after sample gelation to ensure a more isotropic expansion [28]. This process primarily relies on biochemical or physical disruption of the sample and uses methods such as proteinase treatment, surfactant treatment, and heat treatment. We note that the post-expansion sample typically has minimal optical scattering and aberration because the strong scattering features and refractive-index inhomogeneities are largely removed during the homogenization and expansion process.

When disrupting the proteins in the homogenization step, both endogenous biomolecules and molecular tags (e.g., antibodies and fluorescent proteins) may also be fragmented and lost. To better retain these molecules for the subsequent labeling and imaging, target biomolecules and labels are preferably anchored to the hydrogel network. In the original implementation of ExM, molecular tags are anchored to the hydrogel using DNA-conjugated antibodies and fluorescently labeled DNA molecules with a 5′ acrydite moiety [22]. In more recent ExM protocols such as protein retention ExM (proExM), commercially available small-molecule linkers [e.g., Acryloyl-X (AcX)] are used to covalently anchor the proteins to the hydrogel network [29]. Conveniently, the AcX linker can also retain genetically encoded fluorescent proteins and fluorescently labeled antibodies. This compatibility of proExM with conventional immunohistochemistry has thus led to its application to a broad range of samples, including the mammalian and invertebrate brain.

In addition to the molecular linkers mentioned above, a broad range of bioconjugation strategies, including the use of other protein linkers (e.g., methacrylic acid N-hydroxysuccinimide ester (MA-NHS) [30], glutaraldehyde (GA) [30–32], and formaldehyde/acrylamide [33]), nucleic acid linkers (e.g., LableX [34] and MelphaX [35]), membrane labels [36–39], and trifunctional and metabolic linkers [40–42], have been developed to efficiently retain various molecular targets (e.g., DNA, RNA, and biological membranes) to the hydrogel (*see* [23] for a more in-depth discussion on the molecular linkers and conjugation chemistries of ExM). These advances have made ExM an ideal platform for high-resolution, multiplexed mapping of molecular profiles in intact cells and tissue samples.

In summary, ExM is an accessible nanoscale imaging method that takes advantage of physical magnification of the sample and is compatible with conventional fluorescence microscopes and a broad range of molecular labeling strategies. In particular, ExM is suitable for fluorescence imaging of tissue samples because the expansion process increases the effective imaging resolution and reduces the optical scattering and aberration inherent to tissue.

1.3 Combining Expansion Microscopy and Lattice Light-Sheet Microscopy

While LLSM is suitable for high-resolution, low-photobleaching volume fluorescence imaging, lattice light-sheet illumination is sensitive to sample aberration and scattering. ExM helps overcome this issue as the sample digestion and expansion steps serve to homogenize the refractive indices and remove major scattering centers across the tissue samples. In addition, by increasing the expansion factor, effective super-resolution imaging can be achieved using diffraction-limited imaging modes of LLSM.

Finally, the high imaging throughput of LLSM is ideal for imaging a physically enlarged tissue sample quickly.

Taking these potential advantages into consideration, we have developed a sample preparation, imaging, and post-imaging processing pipeline to combine LLSM and ExM for nanoscopic imaging of thick tissue samples. Because the dimensions of expanded tissue samples are often larger than the widths of a single lattice light-sheet, we use sample (i.e., stage) or objective scanning to tile multiple subvolumes to fill the entire imaging volume. We then stitch these subvolumes computationally after imaging [43]. Besides the sample and objective tiling, tiling the lattice light-sheet illumination is another approach to scale up the imaging volume. As opposed to the original implementation of LLSM, tiling lattice light-sheets are generated by optically tiling multiple light-sheets along the light-sheet length direction. As a result, the tiled lattice light-sheets drastically increase the field of view across the illuminated sample [44–46].

Named expansion lattice light-sheet microscopy (ExLLSM or ExL), we have applied this combined imaging modality of LLSM and ExM to the entire mouse cortical column and *Drosophila* brain (Fig. 1) [43]. In the mouse cortex dataset, for example, a ~$280 \times 1900 \times 70$ μm Thyl-YFP mouse primary somatosensory cortex tissue slice (in pre-expansion dimensions) was imaged to capture a sparse subset of cytosolically labeled layer VI-V pyramidal neurons as well as synaptic proteins Bassoon and Homer1 from the pia to the white matter (Fig. 1a). The sub-100 nm resolution and the large dimensions of the dataset allowed us to visualize and reconstruct dendritic spines of the labeled pyramidal neurons across multiple cortical layers (Fig. 1b). In the whole *Drosophila* brain dataset, a ~$340 \times 660 \times 90$ μm volume of an adult *Drosophila* brain (in pre-expansion dimensions) was imaged to capture dopaminergic neurons as well as presynaptic sites across the entirety of the brain (Fig. 1e). The sub-100 nm resolution and the large dimension of the dataset allowed us to assign a subset of the presynaptic sites to the dopaminergic neurons across major anatomical regions of the *Drosophila* brain.

With the ExL pipeline, the intricate morphology and molecular composition of synaptic proteins, myelination sheaths, dendritic spines, and neurites can be imaged at close to ~60 nm effective lateral resolution and ~90 nm effective axial resolution [43]. This spatial resolution can be further improved by increasing the sample expansion factor (*see* [23] for a more in-depth discussion on ExM hydrogel chemistries and expansion factors). Moreover, ExL can be scaled up to large tissue samples (e.g., centimeters or larger in size) by applying additional modifications to the imaging setup (e.g., increasing the range of the sample stage) or by combining it with (iterative) sample sectioning, such as photochemical sectioning [47]. This combination of high spatial resolution and imaging throughput establishes ExL as an ideal platform to study, for example, stereotypical neuronal projections and synaptic connections across multiple brains as well as spatial distributions of protein markers at different developmental and disease states. Recent studies have also used ExL to study the correlation between the animals' behavior and the underlying structural and functional neuronal connectivity [45, 48, 49].

2 Materials

2.1 Tissue Preparation

Expansion lattice light-sheet microscopy (ExLLSM or ExL for short) is compatible with a broad range of sample types. For example, mouse brain slices can be expanded and optically cleared using the protein-retention ExM (proExM) protocol with a minor modification (Fig. 4) [29]. Briefly, Thy1-YFP mice (Thy1-YFP-H strain on C57BL/6, transgenic mice expressing cytosolic YFP under Thy1 promoter) were anesthetized with ketamine/xylazine and perfused transcardially with 25 mL of 1× PBS, followed by 25 mL of ice cold 4% paraformaldehyde (PFA) in 1× PBS. The mouse brains were then chemically fixed in 4% PFA at 4 °C for 24 h. Next, the fixed mouse brains were sectioned into coronal slices (~50–250 μm in thickness) using a vibratome (Leica VT1000S) and stored in 1× PBS at 4 °C (low concentration sodium azide is needed for long-term storage). For immunohistochemistry, the mouse brain slices were permeabilized with 0.1% Triton X-100 in 1× PBS at room temperature for 15 min. To reduce nonspecific binding, a blocking buffer (5% normal donkey serum and 0.1% Triton X-100 in 1× PBS) was applied to the sample at room temperature overnight. The brain slices were then incubated with primary antibodies at a dilution of 1:100 to 1:200 in the blocking buffer on an orbital shaker at room temperature for 24 h before the antibody solution was washed out 4 × 30 min with fresh blocking buffer. Next, the brain slices were incubated with secondary antibodies at a dilution of 1:100 to 1:200 in the blocking buffer on an orbital shaker at room temperature for 24 h before the antibody solution was washed out 4 × 30 min with fresh 1× PBS for the subsequent gelation.

ExL is compatible with clinical samples, such as human kidney slices (US BioMAX, HuFTS301). Briefly, the human tissue slice was immunostained following the expansion pathology (ExPath) protocol [49]. First, the tissue slice was placed in a heat-resistant container with 20 mM sodium citrate solution (pH 8) at ~100 °C, and the container was immediately transferred to a 60 °C incubator. After 30 min incubation, the sample was immunostained with primary antibodies in MAXbind™ Staining Medium (Active Motif) at a concentration of 10 μg/mL at room temperature for 3 h, followed by staining with secondary antibodies (in the same

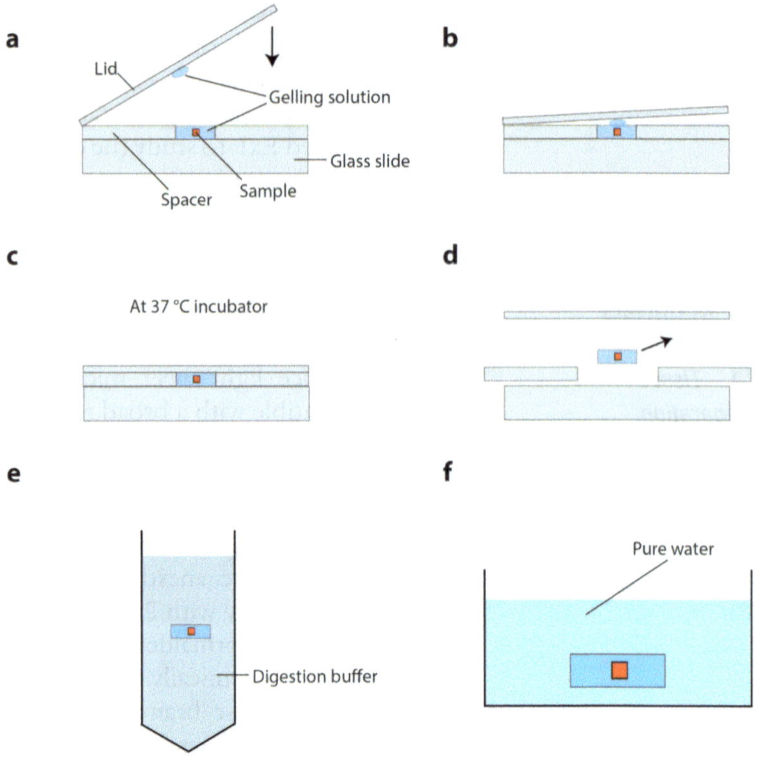

Fig. 4 Gelation and expansion setups for ExL sample preparation. (**a**) A gelation chamber is assembled with two spacers (e.g., coverslips) and a glass slide. The tissue sample as well as the gelling solution are first placed in between the spacers, which are thicker than the tissue. (**b**) The lid is then moved towards the sample. A droplet of gelling solution in the lid can merge with the gelling solution in the chamber, preventing formation of bubbles during the subsequent gelation. (**c**) A correctly assembled gelation chamber. Gelation is performed at 37 °C for 2 h. (**d**) The lid and the spacers are removed after 2 h of gelation. The sample–hydrogel composite can then be trimmed with a razor blade. (**e**) The trimmed gel is placed in the digestion buffer for homogenization. (**f**) The sample–hydrogel composite is typically slightly expanded after the homogenization step

buffer). Finally, the sample was washed two to three times with MAXwash™ Washing Medium to remove the unbound antibodies and dyes before the expansion part of the ExPath protocol.

ExL is also compatible with whole-mount organs and tissue blocks. For example, *Drosophila* brains can be dissected and stained using a previously described histology protocol with a minor modification [50, 51]. Briefly, brains of 3–10-day-old adult female flies were dissected in Schneider's insect medium and chemically fixed in 4% PFA in Schneider's medium for 55 min at room temperature. To help antibodies penetrate the tissue, the fly brains were washed with PBT buffer (0.5% Triton X-100 in 1× PBS) for 3 × 10 min before being blocked in 5% normal goat serum (or normal donkey serum, depending on the secondary antibody) for 90 min. Next, the fly brains were incubated in primary antibodies diluted in 5% serum in PBT for 2–4 days on a nutating shaker at 4 °C, washed in PBT for 3 × 30 min or longer, and incubated in secondary antibodies diluted in 5% serum in PBT for 2–4 days on a nutating shaker at 4 °C.

2.2 Tissue Expansion

Unless otherwise noted, we performed tissue gelation and expansion for ExL using the proExM protocol with a minor modification (Fig. 4) [28, 29, 43]. Briefly, stock solutions including 10 mg/mL acryloyl-X, SE [6-((acryloyl)amino)hexanoic acid, succinimidyl ester or AcX for short, Thermo Fisher] in anhydrous DMSO, 5 M NaCl, 38% (w/v) sodium acrylate, 50% (w/v) acrylamide, 2% (w/v) *N,N'*-methylenebisacrylamide (Bis), 10% (w/v) ammonium persulfate (APS), 10% (w/v) tetramethylethylenediamine (TEMED), and 0.5% (w/v) 4-hydroxy-2,2,6,6-tetramethylpiperidin-1-oxyl (4HT) were prepared and stored at −20 °C in a desiccated environment. Next, fixed and immunostained tissue samples (100 μm or thinner) were incubated with the AcX stock solution at a dilution of 1:100 (0.1 mg/mL in 1× PBS) at room temperature overnight. Monomer solution [2 M NaCl, 8.625% (w/v) sodium acrylate, 2.5% (w/v) acrylamide, 0.15% (w/v) *N,N'*--methylenebisacrylamide)] was prepared and cooled to 4 °C in advance. Specifically, 2.25 mL sodium acrylate stock solution, 0.5 mL acrylamide, 0.75 mL bis, 4 mL sodium chloride, 1 mL 10× PBS, and 0.9 mL water were mixed to prepare 9.4 mL of the monomer solution. Stock solutions of APS, TEMED, and 4HT were then added to the monomer solution at a 1:1:1:47 ratio to a final concentration of 0.2% (w/v), 0.2% (w/v), and 0.01% (w/v), respectively, to yield the gelling solution. In some cases (especially for whole *Drosophila* brains and thick slices, see below), the ratio of 4HT was increased 1.5-fold or more to reduce the risk of premature gelation. To further avoid premature gelation, the gelling solution was temporarily stored at 4 °C or below before quickly moving to the sample gelation step. Next, the tissue samples in the AcX solution were washed 2 × 15 min with 1× PBS and incubated in the gelling solution at 4 °C for 30 min, transferred to a gelation chamber, and gelled in a humidified 37 °C incubator for 2 h (*see* **Note 1**). The gelled samples were trimmed (*see* **Note 1**) and immersed in the digestion buffer [50 mM Tris–HCl (pH 8), 1 mM EDTA, 0.5% Triton X-100, 1 M NaCl, 8 units/mL Proteinase K (New England Biolabs)] at room temperature overnight. The digested samples were expanded in doubly deionized water or 1 mM NaOH aqueous solution for 3 × 20 min before sample mounting and imaging.

For thick mouse brain slices (>100 μm), more time is needed to diffuse reagents and solutions throughout the sample [28, 29]. Briefly, fixed and immunostained mouse brain slices were incubated in the AcX solution (0.1 mg/mL) in MES buffered saline [150 mM NaCl (pH ~6) in 100 mM MES] at room temperature for 24 h to increase the penetration depth of the anchoring reagent. Next, 1.5 times 4HT (0.015% w/v) was added into the gelling solution to avoid premature gelation during the incubation time. The tissue samples were then incubated in the gelling solution at 4 °C for 45–60 min before gelation (*see* **Note 2**). Digestion of the samples was carried out in the digestion buffer at room temperature for 24 h and then in fresh digestion buffer for another 24 h. *Drosophila* brain samples were prepared using similar steps, except for the AcX stock solution was diluted in 1× PBS instead of the MES buffered saline.

For human kidney slices, we used the ExPath protocol with a minor modification [49]. Compared to the proExM protocol used for mouse brain slices and other samples, two major changes were made in the ExPath protocol. First, the concentration of the monomer solution was decreased [2 M NaCl, 8.625% (w/v) sodium acrylate, 2.5% (w/v) acrylamide, 0.10% (w/v) *N,N'*-methylenebisacrylamide]. Second, the digestion temperature was increased to boost the enzymatic activity. For example, the ExPath samples were incubated in the digestion buffer at 60 °C for 3 h or until the completion of digestion.

2.3 Lattice Light-Sheet Microscopes

We imaged most of the expanded tissue samples in the objective scan mode [17] (where the detection objective and the light-sheet moved together in discrete steps) using an LLSM setup equipped with a 12-mm coverslip sample holder but with its adaptive optics capabilities disabled [19]. We also designed an LLSM setup (similar to the first-generation LLSM setup but optimized for expanded samples) to increase the field of view and allow for long-range fast scanning in the sample scan mode (where the sample was translated continuously parallel to the coverslip surface where it is mounted) [43, 48]. In detail, the latter LLSM setup used two pairs of cylindrical lenses to illuminate a thin stripe on a spatial light modulator (Forth Dimension QXGA with 2048 × 1536 pixels) in order to generate a lattice light-sheet of ~160 μm in width along the y-axis. The sample was scanned in the x-direction ("sample scan mode") using a piezo inertia stage (PI Q522.130) with ~1.5 mm scanning distance. The width and depth of the sample bath was increased to facilitate the tiling of samples over a larger range (up to ±6 mm). A two-camera solution was implemented along with dichroics and emission filters (Semrock Di03-R561-t1-25x36; emission filter on the camera 1 is a Semrock FF01-530/43-25; emission filter on the camera 2 was a Semrock LP02-568RU-25), which, in the fast multicolor acquisition mode, allowed readout of one camera during exposure of the other. We note that the first-generation LLSM setup is also compatible with ExL but may require additional modifications to the system, including adjustments to the sample holder (so that it can accommodate large samples) and the control software (so that it can accommodate 3D tiling during the sample scan [17]).

3 Methods

3.1 Expansion and Mounting of Expanded Tissues

To ensure mechanical stability during LLSM imaging, proper mounting of the expanded tissue on the LLSM sample holder is crucial, especially for long-term imaging of larger tissue samples. Unless otherwise noted, we mounted all the expanded tissues using either the "superglue method" or the "poly-L-lysine method" [28]. First, 12-mm coverslips were cleaned by soaking in 1 M KOH for 30 min, rinsed with double-distilled water three times, stored in 30% (v/v) ethanol, and air-dried before sample mounting. For the "superglue method," a thin layer of superglue was applied to a small area of the cleaned 12-mm coverslip. To enhance adhesion between the sample and coverslip, excessive liquid around the expanded sample was wicked away before the sample was placed on the uncured superglue. After curing the superglue for 20–30 s, a few droplets of double-distilled water were added on top of the sample to keep it hydrated and to wash away any uncured superglue (*see* **Note 3**). At this point, an opaque interface was formed between the gel and the superglue. For the "poly-L-lysine method", a few droplets of 0.1% (w/v) poly-L-lysine aqueous solution were applied to the top surface of the cleaned 12-mm coverslip for 20 min. After the poly-L-lysine modified surface was rinsed with double-distilled water three times and air-dried for 1 h in a clean environment, the same steps as in the "superglue method" were performed to wick away excessive liquid and mount the expanded samples (*see* **Note 3**). A few droplets of doubly deionized water or 1 mM NaOH aqueous solution (to elongate the lifetime of endogenous fluorescent proteins) were applied to the sample to prevent sample drying before imaging.

3.2 Imaging Expanded Tissues on a Lattice Light-Sheet Microscope

For selected samples that required long-term imaging (~days), fluorescent beads were applied and attached to the sample surface for autofocusing purposes. First, the bead solution [amine-terminated FluoSpheres, 0.2 μm, red fluorescent (580/605), Thermo Fisher] was diluted in double-distilled water by a factor of ~10^5–10^6 and ultrasonicated for 10 min before use. The sample was then soaked in diluted fluorescent bead solution for 5 min and rinsed with fresh double-distilled water twice before transferring to the sample chamber.

After attaching the sample holder to the sample chamber, the sample was imaged on an LLSM setup in a bath of double-distilled water or 1 mM NaOH aqueous solution at room temperature. Samples that required long-term imaging (~days) were perfused with fresh double-distilled water or 1 mM NaOH aqueous solution to maintain a stable pH and to prevent denaturing of fluorescent proteins.

3.3 Image Processing and Visualization

After data collection, a large number of 3D image tiles (i.e., subvolumes) need to be stitched together to reconstruct the imaged volume. For this purpose, an image preprocessing pipeline utilizing Apache Spark 2-based high-performance computing (HPC) environments was developed by Stephan Saalfeld and Igor Pisarev to enable rapid flat-field correction, deconvolution, and stitching of datasets with thousands of 3D image tiles [43]. These software packages are available as open-source projects on GitHub (flat-field correction, stitching, and export: https://github.com/saalfeldlab/stitching-spark; visualization tools: https://github.com/saalfeldlab/n5-viewer). All the ExL datasets were preprocessed with this flat-field correction, deconvolution, and stitching pipeline, which was executed on a HPC cluster. In addition, selected ExL datasets were processed with deskewing/rotation, noise removal, or chromatic offset correction.

The flat-field correction pipeline was developed based on CIDRE [52]. The new implementation supports both 2D and 3D images at full resolution from the ExL pipeline, has constant memory requirements, and relaxes the optimization objective in a way that enables parallel solution. For all the ExL datasets acquired, flat-fields were calculated for each dataset and color channel independently. The flat-fields were then applied to all the 3D image tiles prior to deconvolution and stitching.

Second, all the ExL datasets were deconvolved after flat-field correction on a tile-by-tile basis. The deconvolution was performed using the Richardson-Lucy algorithm adapted to run on MATLAB with experimentally measured PSFs for each color channel for 15 iterations [17].

Third, all the ExL datasets were stitched with the aforementioned stitching pipeline of rigid transformation based on image tile cross-correlations, which incorporated an iteratively refined prediction model for systematic stage coordinate errors. The fully automated prediction model enabled stitching of all datasets without manual intervention [43].

Finally, the stitched volume was exported to a custom file format (N5) that enabled parallel block-wise export and compression as a multichannel multiscale pyramid using a parallel HPC exporter [43]. A Fiji plugin, "N5 Viewer," was developed based on BigDataViewer [53] for interactive browsing of the multichannel multiscale N5 datasets [43]. The plugin automatically saves user adjusted viewer properties such as color scheme, contrast, grouping, and bookmarks to simplify working with datasets and sharing them with colleagues. Arbitrary subvolumes were extracted and exported for processing in Fiji or other processing pipelines.

The data collected using the sample scan mode were deskewed with inverse shear transform [17, 54]. The deskewed sample-scan data and selected objective scan datasets were corrected for Z-anisotropy by 3D interpolation and rotated along the Y axis and into the real-world coordinates. In order to decrease the overall file sizes and facilitate visualization, the rotated data was anisotropically resampled in X and Z dimensions to match the original Nyquist sampling of the collected data.

3.4 Reconstruction and Analysis of Subcellular Structures

To analyze the morphology and molecular profile of subcellular structures within the ExL datasets, a number of commercial and custom computational and analysis pipelines were used to, for example, segment the organelles, calculate the G-ratio of axons, and reconstruct the dendritic spines [43]. However, despite such developments, custom code development is still necessary to address specific biological questions.

For volumetric rendering, a combination of commercial and open-source software, such as Vision4D (Arivis), Imaris ×64 8.4–9.1 (Oxford Instruments), VVD Viewer (ver. 1.00, HHMI Janelia Research Campus), and Amira 6.3–6.5 (Thermo Fisher) were used.

4 Conclusion

In this chapter, we have demonstrated the general principle and workflow of ExL and its capability to capture the sub-diffraction-limit anatomy and molecular profile across large brain tissue samples (e.g., samples of ~millimeters in size). We have also described the technical requirements and caveats to implement the standard ExL imaging pipeline.

For ExL sample preparation, we have primarily focused on ~4× expansion of mouse brain slices and insect brains. As described earlier, samples with larger expansion can be prepared by tuning the expansion hydrogel chemistry [23]. However, additional factors need to be considered before preparing and imaging samples of higher expansion factors for ExL. First, the sample thickness increases drastically with the expansion. Objectives with longer working distances are thus needed to image through the expanded tissue samples with higher expansion factors [55]. However, increasing the objective working distance often results in the reduction of numerical aperture (NA), which can lead to a decrease in the effective resolution and an increase in the photobleaching effect. This issue is addressable by iteratively applying spatially precise and gentle sample sectioning methods, such as photochemical sectioning, and on-block imaging [47]. Second, expanded tissue samples of higher expansion factors encounter the same labeling issue that other high-resolution fluorescence microscopy samples face. As the effective resolution increases (i.e., the sample expansion factor increases), the volumetric density of fluorescent labels decreases drastically. This dilution effect of the labeling can lead to additional imaging artifacts and prevent reconstruction of sparse and dim molecular targets in the expanded tissue samples.

Taken together, these factors point to a number of future developments that can help enhance the resolution and scalability of ExL. First, serial sample sectioning (e.g., photochemical sectioning) or the use of objectives with high NA and ultra-long working distance are needed to scale up ExL with increasingly larger sample dimensions. Second, a ubiquitous, dense, and cell-permeable fluorescent stain is desired to increase the sample labeling density. Finally, as the target resolution and sample size increase, a robust and scalable computing and data storage infrastructure is necessary to flat-field correct, stitch, export, and analyze the increasingly large ExL datasets.

5 Notes

1. For brain tissue gelation and expansion, we find that a few tips can help increase the success rate. First, bubbles formed inside the gel can cause sample damage, anisotropic expansion, and additional aberration during imaging. To mitigate this issue, it is often useful to spin the gelling solution down on a desk-top centrifuge for 2–3 min before applying it to the sample and fill the gelation chamber. Second, a humidified gelation environment can be created by simply putting a piece of crumpled wet paper towel in a closed petri dish. When a wet paper towel is used, make sure that it is not in touch with any assembled gelation chambers because the wet paper towel can wick away the gelling solution over time. Third, after gelation, we suggest that the sample be trimmed into an asymmetric shape. As a result, one can easily find the side of the gel to which biological samples are closer (note that due to gravity, tissues tend to be gelled closer to the bottom side of the gel in the gelation chamber) and image deeper into the sample.
2. A larger amount of gelling solution is used for thick tissue samples. Therefore, it is important to prevent the gelation solution from leaking during the gelation process. We often cover the bottom of the petri dish with parafilm before placing the assembled gelation chambers.
3. After mounting the expanded samples on the 12-mm coverslip, we suggest gently washing the samples two to three times by applying droplets of double-distilled water or 1 mM NaOH aqueous solution and sucking them away, regardless of which mounting method is used. This is because additional

contaminants can be introduced during the mounting process. Sample detachment can happen during a long imaging session (of multiple days). If the sample mounted with the "poly-L-lysine method" gets detached, the sample can be remounted onto another freshly prepared poly-L-lysin modified coverslip. For the "superglue method," however, the mounting is more or less irreversible. Therefore, a proper superglue (e.g., gel type vs. liquid type) should be tested before sample mounting.

Conflict of Interest

R.G. is a co-inventor of multiple patents related to expansion microscopy. The other authors have no conflicts of interest to declare that are relevant to the content of this chapter.

References

1. Betzig E, Patterson GH, Sougrat R, Lindwasser OW, Olenych S, Bonifacino JS, Davidson MW, Lippincott-Schwartz J, Hess HF (2006) Imaging intracellular fluorescent proteins at nanometer resolution. Science (1979) 313: 1642–1645
2. Heilemann M, Van De Linde S, Schüttpelz M, Kasper R, Seefeldt B, Mukherjee A, Tinnefeld P, Sauer M (2008) Subdiffraction-resolution fluorescence imaging with conventional fluorescent probes. Angew Chem Int Ed 47:6172–6176
3. Rust MJ, Bates M, Zhuang X (2006) Sub-diffraction-limit imaging by stochastic optical reconstruction microscopy (STORM). Nat Methods 3:793
4. Klar TA, Jakobs S, Dyba M, Egner A, Hell SW (2000) Fluorescence microscopy with diffraction resolution barrier broken by stimulated emission. Proc Natl Acad Sci USA 97:8206
5. Pelicci S, Tortarolo G, Vicidomini G, Diaspro A, Lanzanò L (2020) Improving SPLIT-STED super-resolution imaging with tunable depletion and excitation power. J Phys D Appl Phys 53:234003
6. Moffitt JR, Osseforth C, Michaelis J (2011) Time-gating improves the spatial resolution of STED microscopy. Opt Express 19:4242
7. Keller J, Schönle A, Hell SW (2007) Efficient fluorescence inhibition patterns for RESOLFT microscopy. Opt Express 15:3361
8. Hell SW (2005) Fluorescence nanoscopy: breaking the diffraction barrier by the RESOLFT concept. NanoBiotechnology 1: 296
9. Gustafsson MGL (2000) Surpassing the lateral resolution limit by a factor of two using structured illumination microscopy. J Microsc 198:82
10. Heintzmann R, Huser T (2017) Super-resolution structured illumination microscopy. Chem Rev 117:13890–13908
11. Balzarotti F, Eilers Y, Gwosch KC, Gynnå AH, Westphal V, Stefani FD, Elf J, Hell SW (2017) Nanometer resolution imaging and tracking of fluorescent molecules with minimal photon fluxes. Science (1979) 355:606–612
12. Deguchi T et al (2023) Direct observation of motor protein stepping in living cells using MINFLUX. Science (1979) 379:1010
13. Wolff JO, Scheiderer L, Engelhardt T, Engelhardt J, Matthias J, Hell SW (2023) MINFLUX dissects the unimpeded walking of kinesin-1. Science (1979) 379:1004
14. Santi PA (2011) Light sheet fluorescence microscopy: a review. J Histochem Cytochem 59:129
15. Power RM, Huisken J (2017) A guide to light-sheet fluorescence microscopy for multiscale imaging. Nat Methods 14:360
16. Stelzer EHK, Strobl F, Chang B-J, Preusser F, Preibisch S, McDole K, Fiolka R (2021) Light sheet fluorescence microscopy. Nat Rev Methods Primers 1:73
17. Chen B-C et al (2014) Lattice light-sheet microscopy: imaging molecules to embryos at high resolution. Science (1979) 346:439+
18. Liu G, Ruan X, Milkie DE, Görlitz F, Mueller M, Hercule W, Killilea A, Betzig E, Upadhyayula S (2023) Characterization,

comparison, and optimization of lattice light sheets. Sci Adv 9:eade6623
19. Liu T-L et al (2018) Observing the cell in its native state: imaging subcellular dynamics in multicellular organisms. Science (1979) 360: eaaq1392
20. McArthur K et al (2018) BAK/BAX macropores facilitate mitochondrial herniation and mtdna efflux during apoptosis. Science (1979) 359:eaao6047
21. Valm AM, Cohen S, Legant WR, Melunis J, Hershberg U, Wait E, Cohen AR, Davidson MW, Betzig E, Lippincott-Schwartz J (2017) Applying systems-level spectral imaging and analysis to reveal the organelle interactome. Nature 546:162
22. Chen F, Tillberg PW, Boyden ES (2015) Expansion microscopy. Science (1979) 347: 543
23. Wang W, Chan YH, Kwon S, Tandukar J, Gao R (2022) Nanoscale fluorescence imaging of biological ultrastructure via molecular anchoring and physical expansion. Nano Converg 9: 30
24. Katchalsky A, Michaeli I (1955) Polyelectrolyte gels in salt solutions. J Polym Sci 15:69
25. Flory PJ (2004) Thermodynamics of high polymer solutions. J Chem Phys 10:51
26. Flory PJ, Rehner JJ (2004) Statistical mechanics of cross-linked polymer networks II. Swelling. J Chem Phys 11:521
27. Hildebrand JH (2004) The entropy of solution of molecules of different size. J Chem Phys 15: 225
28. Asano SM, Gao R, Wassie AT, Tillberg PW, Chen F, Boyden ES (2018) Expansion microscopy: protocols for imaging proteins and RNA in cells and tissues. Curr Protoc Cell Biol 80: e56
29. Tillberg PW et al (2016) Protein-retention expansion microscopy of cells and tissues labeled using standard fluorescent proteins and antibodies. Nat Biotechnol 34:987
30. Chozinski TJ, Halpern AR, Okawa H, Kim HJ, Tremel GJ, Wong ROL, Vaughan JC (2016) Expansion microscopy with conventional antibodies and fluorescent proteins. Nat Methods 13:485
31. Chen L, Yao L, Zhang L, Fei Y, Mi L, Ma J (2021) Applications of super resolution expansion microscopy in yeast. Front Phys 9:650353
32. Goetz R, Panzer S, Trinks N, Eilts J, Wagener J, Turra D, Di Pietro A, Sauer M, Terpitz U (2020) Expansion microscopy for cell biology analysis in fungi. Front Microbiol 11:574
33. Ku T, Swaney J, Park JY, Albanese A, Murray E, Hun Cho J, Park YG, Mangena V, Chen J, Chung K (2016) Multiplexed and scalable super-resolution imaging of three-dimensional protein localization in size-adjustable tissues. Nat Biotechnol 34:973
34. Chen F et al (2016) Nanoscale imaging of RNA with expansion microscopy. Nat Methods 13:679
35. Wang Y et al (2021) EASI-FISH for thick tissue defines lateral hypothalamus spatio-molecular organization. Cell 184:6361
36. Götz R, Kunz TC, Fink J, Solger F, Schlegel J, Seibel J, Kozjak-Pavlovic V, Rudel T, Sauer M (2020) Nanoscale imaging of bacterial infections by sphingolipid expansion microscopy. Nat Commun 11:6173
37. Karagiannis ED, Kang JS, Shin TW, Emenari A, Asano S, Lin L, Costa EK, Boyden ES, Marblestone AH, Kasthuri N (2019) Expansion microscopy of lipid membranes. BioRxiv
38. Wen G, Vanheusden M, Acke A, Valli D, Neely RK, Leen V, Hofkens J (2020) Evaluation of direct grafting strategies via trivalent anchoring for enabling lipid membrane and cytoskeleton staining in expansion. ACS Nano 14:7860
39. White BM, Kumar P, Conwell AN, Wu K, Baskin JM (2022) Lipid expansion microscopy. J Am Chem Soc 144:18212
40. Sun D et al (2021) Click-ExM enables expansion microscopy for all biomolecules. Nat Methods 18:107
41. Shi X et al (2021) Label-retention expansion microscopy. J Cell Biol 220:e202105067
42. Wen G, Vanheusden M, Leen V, Rohand T, Vandereyken K, Voet T, Hofkens J (2021) A universal labeling strategy for nucleic acids in expansion microscopy. J Am Chem Soc 143: 13782
43. Gao R et al (2019) Cortical column and whole-brain imaging with molecular contrast and nanoscale resolution. Science (1979) 363: eaau8302
44. Gao L (2015) Extend the field of view of selective plan illumination microscopy by tiling the excitation light sheet. Opt Express 23:6102
45. Gao L, Tang W-C, Tsai Y-C, Chen B-C (2019) Lattice light sheet microscopy using tiling lattice light sheets. Opt Express 27:1497
46. Tsai Y-C et al (2020) Rapid high resolution 3D imaging of expanded biological specimens with lattice light sheet microscopy. Methods 174:11
47. Wang W et al (2025) Mesoscale volumetric fluorescence imaging at nanoscale resolution by photochemical sectioning. Science 390(6770):eadr9109. https://doi.org/10.1126/science.adr9109

48. Lillvis JL et al (2022) Rapid reconstruction of neural circuits using tissue expansion and light microscopy. Elife 11:e81248
49. Zhao Y et al (2017) Nanoscale imaging of clinical specimens using pathology-optimized expansion microscopy. Nat Biotechnol 35:757
50. Nern A, Pfeiffer BD, Rubin GM (2015) Optimized tools for multicolor stochastic labeling reveal diverse cell arrangements in the fly visual system. Proc Natl Acad Sci USA 112:E2967
51. Jenett A et al (2012) A GAL4-driver line resource for drosophila neurobiology. Cell Rep 2:991
52. Smith K, Li Y, Piccinini F, Csucs G, Balazs C, Bevilacqua A, Horvath P (2015) CIDRE: an illumination-correction method for optical microscopy. Nat Methods 12:404+
53. Pietzsch T, Saalfeld S, Preibisch S, Tomancak P (2015) BigDataViewer: visualization and processing for large image data sets. Nat Methods 12:481
54. Aguet F et al (2016) Membrane dynamics of dividing cells imaged by lattice light-sheet. Mol Biol Cell 27:3418
55. Glaser A et al (2025) Expansion-assisted selective plane illumination microscopy for nanoscale imaging of centimeter-scale tissues. eLife 12:RP91979. https://doi.org/10.7554/eLife.91979.3

Chapter 6

A Guide to Leveraging Electronics Metrology Technologies for New Light-Sheet Microscopy Systems

Adam Glaser

Abstract

Light-sheet microscopy has emerged as the technique of choice for fast in vivo and ex vivo imaging of biological systems and processes. The selection of the objective lens and camera is a critical decision that dictates the resolution, field of view, working distance, imaging speed, and overall capability of any light-sheet microscopy system. Historically, nearly every light-sheet microscope has selected these two technologies from the life sciences domain. However, there are extremely powerful technologies that have been developed by the adjacent electronics metrology industry. Here, I summarize and discuss the available technologies from this industry and how they can be leveraged to design new, powerful light-sheet microscopy systems.

Key words Light-sheet microscopy, Electronics metrology, Etendue, Optical design

1 Introduction

Within the life sciences, optical microscopes enable the study of biological specimens with fine detail. Images captured by traditional microscope objective lenses exhibit trade-offs between the smallest objects that can be resolved (i.e., the resolution), how much of the specimen can be observed (i.e., the field of view), and how thick a specimen may be (i.e., the working distance). While these trade-offs are not inherent physical laws, they are limitations imposed by the practical limits of optical design, engineering, and lens manufacturing.

The development of microscope objective lenses has a long history spanning over hundreds of years. Present-day commercial microscopes have become highly standardized and strongly application-oriented, with objective lenses designed specifically around the technology roadmap of the given microscope vendor. For example, the resolution, field of view, and working distance of lenses has been guided by the demands of pre-clinical and clinical applications that require the observation of individual cells, tissues,

Ludovico Silvestri (ed.), *Light Sheet Microscopy*, Neuromethods, vol. 229, https://doi.org/10.1007/978-1-0716-5100-1_6,

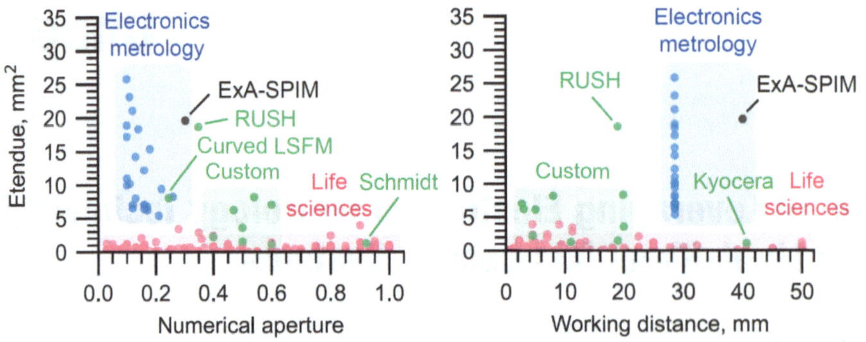

Fig. 1 Etendue as a function of numerical aperture (left) and working distance (right) for electronics metrology lenses (blue), life sciences lenses (magenta), and custom lenses (green)

and organisms. This has resulted in an ecosystem of lenses across all life sciences microscope vendors with similar trade-offs between resolution, field of view, and working distance.

The comparative metric which describes the trade-off between resolution and field of view is known as the etendue (G), defined as a function of the lens field of view (FOV) and numerical aperture (NA) as:

$$G = \frac{\pi}{4}(\mathrm{FOV} \times \mathrm{NA})^2$$

In a recent review, Zhang and Gross performed a comprehensive historical review of 448 objective lens patents granted in the United States, Germany, and Japan between 1926 and 2008 [1]. Except for 38 objective lenses, their analysis revealed that the etendue of >90% of microscope objective lenses is bounded by $G = 0.0243\ \mathrm{mm}^2$ and $G = 0.9503\ \mathrm{mm}^2$ (Fig. 1), with the maximum etendue of $G = 3.976\ \mathrm{mm}^2$ for an objective lens with a FOV = 2.5 mm and NA = 0.90. Assignees for these patents include Olympus, Nikon, Leica, and Carl Zeiss, the four major vendors of microscope objectives, as well as smaller vendors including American Optical Corporation and Mitutoyo.

A custom academically developed lens known as the Mesolens is aimed at breaking this trade-off and providing an unprecedented etendue [2]. It has a reported etendue of $G = 6.245\ \mathrm{mm}^2$ with a FOV = 6 mm and NA = 0.47. However, due to the practical limitations of engineering and manufacturing such a lens, there has been no dissemination or widespread adoption of the Mesolens [3]. More recently, additional lenses summarized in Table 1 have been developed.

Table 1
Summary of custom lenses developed for life sciences microscopy

References	Name	Immersion	*f*	NA	FOV (mm)	*G* (mm^2)	Working distance (mm)
Tang et al. [4]	Curved LSFM	1.33–1.56	118	0.25	13.00	8.30	20.00
Fan et al. [5]	RUSH	1.33–1.56	–	0.35	14.00	18.86	19.0
Keisuke Ota et al. [6]	FASHIO-2PM	–	35	0.40	4.24	2.26	4.50
McConnel et al. [2]	Mesolens	–	50	0.47	6.00	1.13	3.00
Yu et al. [7]	Cousa	1.33–1.56	20	0.50	2.00	0.79	20.00
Yu et al. [8]	Diesel2P		30	0.54	6.00	8.24	8.00
Kyocera	CS06-10-40-154	1.52–1.56	18	0.60	2.00	1.13	40.60
Sofroniew et al. [9]	2p-RAM	–	21	0.60	5.00	7.07	2.70
Voigt et al. [10]	Schmidt	1.33–1.56	–	1.00	1.15	1.04	11.00

2 Large Etendue Machine Vision Lenses

Rather than designing a highly customized optical lens from scratch (i.e., the Mesolens) it would be extremely advantageous to leverage engineering investments into lenses with similar capabilities from a parallel domain.

High-resolution and high-throughput imaging has become paramount within the machine vision and electronics inspection industry. As opposed to observing biological specimens, optical lenses are instead used to inspect and identify defects in electronics and semiconductor devices. Similarly, the objective lenses used in this domain have become standardized and developed accordingly with the demands of the given inspection application. Over the last several decades, the physical size of components within electronics (e.g., pixels on flat panel displays) has become exceedingly smaller, approaching a biologically relevant length scale (i.e., 1–10 μm). As a result, lenses for this domain are being designed with increasingly large NA. While most lenses within this domain still have NA <0.1, more recent lenses with a NA >0.1 have been released.

A non-comprehensive list of the lenses with NA >0.1 and the highest etendues (maximum $G = 19.651$ mm^2) are shown in Table 2. All of these lenses exceed the etendue bounds of the Mesolens, as well as all objective lenses analyzed by Zhang and Gross [1]. Interestingly, to the best of our knowledge, there have been no investigations or demonstrations of using these lenses for

Table 2
List of high-etendue optical lenses from the machine vision domain

Vendor	Lens model	NA	FOV (mm)	G (mm^2)
Excelitas	d.fine HR 2.5X / 128	0.13	24.62	8.05
Schneider	Xenon Diamond 5.0X / F1.5	0.22	15.77	9.45
Schneider	Xenon Diamond 3.3X / F2.2	0.17	23.43	12.46
Schneider	Xenon Diamond 2.5X / F2.7	0.13	31.54	13.20
Vieworks	VEO_JM DIAMOND 5.0X / F1.3	0.31	16.40	19.65
Vieworks	VEO_JM DIAMOND 3.33X / F2.1	0.18	24.62	15.43
Vieworks	VEO_JM DIAMOND 2.5X / F2.6	0.13	32.80	15.17

life sciences applications. This is despite the extraordinary specifications and etendue values that these lenses provide, when compared to the objective lenses designed specifically for the life sciences applications. Furthermore, unlike the Mesolens, these lenses are readily available and are manufactured in large quantities for the electronics inspection industry. Therefore, they are extremely cost-effective and easily disseminated as compared to bespoke objective lens designs. This is due to the drastic difference in market size between life sciences microscopy, estimated to reach $1.6 billion by 2027, and the machine vision systems, estimated to reach $74.9 billion by 2027 [11].

In addition to an extremely high etendue, these lenses are also extremely well-suited for biological applications because they exhibit:

1. *Low field curvature*: It is notoriously difficult to design lenses which have low field curvature over the specified FOV. Within the machine vision inspection industry, low field curvature is required to inspect electronics which are extremely flat. As a result, these lenses have been designed with extremely flat imaging fields. This is extremely advantageous for biological applications, where the accurate and quantitative observation of biological specimens also requires a flat imaging field.
2. *Low distortion*: In addition to out-of-plane curvature, all lenses also exhibit in-plane curvature or distortion. This distortion is undesirable, as it can hinder downstream quantification and processing of images. Fortunately, lenses for electronics inspection are designed with extremely low distortion to enable accurate quantification and localization of manufacturing defects. This is advantageous for biological applications, where nonlinear distortion of features at the edges of the imaging field can make alignment and stitching of panoramic images challenging.

3. *Chromatic correction*: When designing objective lenses, performance must also be balanced over a given chromatic range. It is desirable for lenses to be corrected over as large a range as possible, as a lens compatible with a limited and narrow range has reduced utility. Within the machine vision inspection industry, components are inspected with visible white light which spans the spectral wavelength range of ~450–700 nm. This visible wavelength range is ideal for an array of biological applications (i.e., brightfield and fluorescence microscopy).

3 Machine Vision Lenses for Light-Sheet Fluorescence Microscopy

Over the past decade, light-sheet fluorescence microscopy (LSFM) has emerged as the technique of choice for fast and gentle 3D microscopy of relatively transparent specimens. In LSFM, a thin sheet of light illuminates a specimen such that fluorescence is selectively excited within a single optical section. The fluorescence is imaged onto a high-speed camera in the direction perpendicular to the light sheet. The initial LSFM architecture, known as selective-plane illumination microscopy (SPIM), was purposefully designed for imaging small model organisms, often living, over long timescales. However, in recent years, LSFM has also been harnessed for 3D microscopy of larger cleared tissues, where high imaging speed and lower rates of photobleaching are the main advantages over alternative microscopy methods.

To perform 3D imaging, LSFM systems must image deep into biological specimens which exhibit a refractive index >1. To avoid reduced image quality (e.g., due to spherical aberration), this necessitates imaging lenses which are designed for direct immersion into a liquid medium which matches the refractive index of the biological specimen. Within the life sciences, there are very few options for lenses to perform this type of imaging. Apart from a handful of lenses, options in this category have extremely short working distances which are not practical for 3D LSFM imaging. A summary of these lenses is shown in Fig. 2, adapted from [12].

Incredibly, machine vision lenses exhibit another unique property which can be leveraged for aberration-free imaging of non-air-based specimens. Most machine vision lenses, including those summarized in Table 2, are designed for co-axial illumination. This optical scheme involves the addition of a glass beam splitter onto the front object side of the lens. The beam splitter enables white light to be coupled into the lens for bright illumination of electronic components during inspection. When used for 3D LSFM imaging, the beam splitter can be removed and replaced with an optically equivalent thickness of the liquid RI media that the biological specimen is immersed within. This scheme avoids the

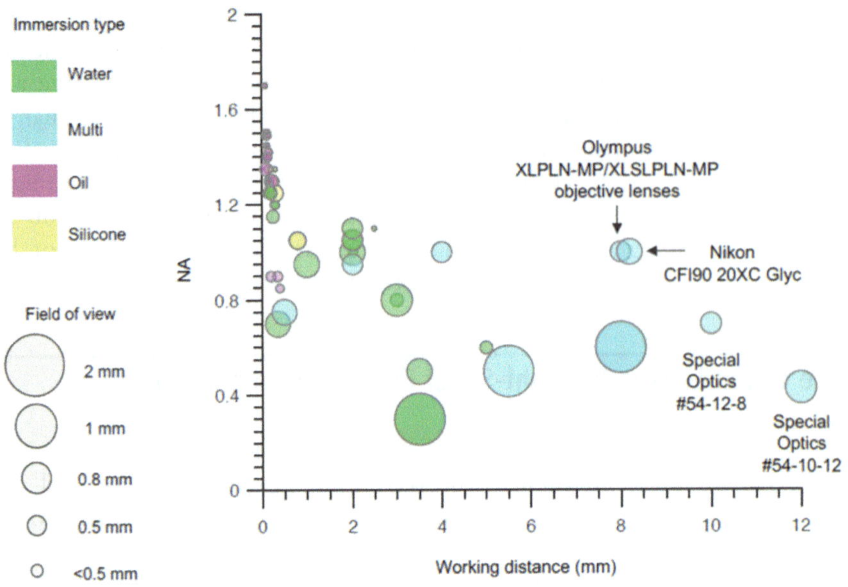

Fig. 2 Summary of life sciences objective lenses designed for immersion in non-air-based media, as required for 3D LSFM. The best existing lenses in terms of FOV, NA, and WD are highlighted

Table 3
Beamsplitter thickness and working distance for lenses listed in Table 2

Vendor	Lens model	BS (mm)	WD (mm)
Excelitas	d.fine HR 2.5X / 128	25	28
Schneider	Xenon Diamond 5.0X / F1.5	25	28
Schneider	Xenon Diamond 3.3X / F2.2	25	28
Schneider	Xenon Diamond 2.5X / F2.7	25	28
Vieworks	VEO_JM DIAMOND 5.0X / F1.3	35	40
Vieworks	VEO_JM DIAMOND 3.33X / F2.1	25	28
Vieworks	VEO_JM DIAMOND 2.5X / F2.6	25	28

severe spherical aberrations that would otherwise plague a standard air-based lens which is used to image into a non-air-based medium.

While the physical working distance of these machine visions is extremely long, the effective working distance for 3D LSFM imaging becomes this optically equivalent thickness. The glass beam splitter thickness (BS) and the optically equivalent thicknesses for imaging into water (WD) for the lenses in Table 2 are summarized in Table 3 when imaging into water (RI = 1.33) for lenses designed for BK7 glass beam splitters (RI = 1.52).

Table 4
Volumetric etendue values for various life sciences and machine vision lenses

Vendor	Lens model	WD (mm)	G^* (mm^3)
Nikon	CFI90 20X GLYC	8	6.28
Olympus	XLPLN-MP	8	6.28
Special Optics	54-10-12	12	2.16
Special Optics	54-12-8	10	0.69
N/A	Mesolens	3	8.81
Excelitas	d.fine HR 2.5X / 128	28	29.30
Schneider	Xenon Diamond 5.0X / F1.5	28	58.23
Schneider	Xenon Diamond 3.3X / F2.2	28	59.30
Schneider	Xenon Diamond 2.5X / F2.7	28	48.06
Vieworks	VEO_JM DIAMOND 5.0X / F1.3	40	239.74
Vieworks	VEO_JM DIAMOND 3.33X / F2.1	28	77.77
Vieworks	VEO_JM DIAMOND 2.5X / F2.6	28	56.93

$$\mathrm{WD}_1 = \frac{\mathrm{RI}_0}{\mathrm{RI}_1}\mathrm{WD}_0$$

Because LSFM systems decouple the illumination and collection paths of the microscope, it is possible to achieve isotropic resolution using a variety of approaches [13, 14]. Therefore, we can calculate a volumetric etendue of a given lens by assuming isotropic resolution over the given working distance of the lens. We define this volumetric etendue (G^*) as:

$$G^* = \frac{\pi}{4}(\mathrm{FOV} \times \mathrm{NA})^2(\mathrm{WD} \times \mathrm{NA})$$

We can calculate the volumetric etendue for a variety of lenses, including the Mesolens, the specialized immersion lenses highlighted in Fig. 2, and the machine lenses listed in Table 2. As can be seen from Table 4, the machine vision lenses possess a volumetric etendue which is orders of magnitude larger than existing life sciences lenses for 3D LSFM imaging.

4 Designing a Light-Sheet Microscope with Machine Vision Lenses

4.1 Detection Lens

From the previous sections, the optimal high-etendue machine vision lens is the VEO_JM DIAMOND 5.0X / F1.3 lens from Schneider-Kreuznach. The detailed specifications for the lens are shown in Table 5. This particular lens offers a volumetric etendue

Table 5
Summary of specifications for the VEO_JM DIAMOND 5.0X / F1.3 lens

VEO_JM DIAMOND 5.0X / F1.3	
Magnification	5.0
NA	0.305
FOV (mm)	16.4
WD (mm)	40
Wavelength (nm)	400–700
Distortion (%)	<0.1
Vignetting (%)	<5
G (mm^2)	19.65
G^* (mm^3)	239.74

which is roughly two orders of magnitude larger than any life sciences lens, and an order of magnitude larger than any other machine vision lens.

This lens forms the basis of a new large-scale light-sheet fluorescence microscope. Based on this lens selection, an appropriate camera is also needed.

4.2 Camera

Nearly every academic and commercial LSFM system to date has used a high-speed scientific CMOS camera for data collection. Historically, these sensors featured an array of 2048 × 2048 pixels with a pitch of 6.5 μm. These specifications are well-matched to the etendue of life sciences objective lenses. However, they are not well-suited for machine vision lenses, which are finite-conjugate and provide a fixed magnification.

This contrasts with modern life sciences objective lenses which are infinity-corrected, with a magnification which can be adjusted by varying the tube lens focal length or introducing various post-magnification lenses. Machine vision lenses must be carefully paired with an appropriate digital camera, where the lens magnification and numerical aperture are matched to the individual pixel size, and the lens field of view is matched to the overall sensor size. The Diamond series lenses in Tables 2 and 3 are highlighted in the upper right portion of the plot.

However, an array of traditional CMOS sensors can be selected which closely match the desired sensor characteristics for high-etendue machine vision lenses. A list of candidate sensors with >100 megapixels is summarized in Table 6. Of importance for 3D LSFM, the large-format Sony sensors provide a single rolling shutter (that rolls from one end of the chip to the other). This is in contrast to a dual rolling shutter, which features two shutters which

Table 6
List of high megapixel sensors for pairing with high-etendue lenses

Vendor	Sensor	Megapixels	Pixel size (μm)	Shutter type
Sony	IMX411	151	3.76	Rolling (single)
Sony	IMX461	100	3.76	Rolling (single)
Sony	IMX661	129	3.45	Global
Samsung	ISOCELL HM3	108	0.80	Rolling (dual)
Samsung	ISOCELL HP1	200	0.64	Rolling (dual)
Canon	120MXS	120	2.20	Rolling (dual)
Canon	LI8020S	250	1.50	Rolling (dual)
Gpixel	GMAX32103	104	3.20	Global
Gpixel	GMAX32152	152	3.20	Global

start at the center of the chip and roll in opposite directions. The single rolling shutter can be leveraged for the so-called axial-sweeping, which is a light-sheet illumination strategy capable of providing near-isotropic or even isotropic resolution. Based on these criteria, the Sony IMX sensors are optimal for large-scale light-sheet microscopy. In particular, the IMX411 and IMX461 sensors provide a single-sided rolling shutter that is compatible with techniques such as axial sweeping and digital scanning [13, 15]. Similar to scientific CMOS sensors, current large-format CMOS sensors (including the IMX sensors) are capable of streaming data at rates exceeding 1 gigapixel/s.

4.3 Excitation Lens

An appropriate light-sheet excitation lens is needed to pair with the high-etendue collection lens. Desired specifications include (1) infinity correction to enable axial sweeping at a conjugated plane, (2) a field of view capable of generating a light-sheet wide enough to cover the entire imaging field of view, (3) NA as close as possible to the collection lens NA for near-isotropic imaging, and (4) a long working distance to enable large-volume imaging. Based on these criteria, there are only several possible options that are listed in (Table 7). To our knowledge, no other existing lenses (e.g., telecentric scan or f-theta lenses) can meet these three requirements, although custom versions of these lenses that satisfy these requirements would be possible to design and fabricate.

4.4 Expansion-Assisted Selective Plane Illumination Microscopy (ExA-SPIM)

It has recently been demonstrated that the ExA-SPIM microscope leverages a unique combination of machine vision lenses and camera sensors for unprecedented volumetric imaging performance of large intact expanded tissues [16]. ExA-SPIM has a field of view ~100× larger (10.6×8.0 mm^2) and a working distance ~10× larger (35 mm) compared to microscope objectives typically used for

Table 7
List of compatible excitation lenses

Vendor	Lens model	NA	FOV (mm)	WD (mm)
Olympus	XLFLUOR-2X	0.14	11	21.0
Thorlabs	TL2X-SAP	0.10	11	56.3
Navitar	1-290419	0.133	16	39.0

Table 8
ExA-SPIM imaging specifications

System specifications	
Lateral resolution (μm)	1.0
Axial resolution (μm)	3.0
Imaging volume (mm^3)	200 × 52 × 35
Wavelength range (nm)	400–700
Refractive index compatibility	1.33–1.56
Imaging speed (voxels/sec)	Up to 946×10^6

biological microscopy, while retaining a relatively high numerical aperture (NA) = 0.305 (resolution of ~1 μm). When combined with 4× tissue expansion, the system achieves a near-isotropic (i.e., the axial resolution is <3× the lateral resolution) effective optical resolution of ~250 nm laterally and ~750 nm axially, at imaging speeds up to 10× faster (1000 megavoxels/s) than the majority of current imaging systems. These unprecedented capabilities allow ExA-SPIM to simultaneously see both the large (centimeter scale tissues, including mouse brains) and small (nanoscale biological structure). The imaging specifications of the ExA-SPIM are summarized in Table 8. ExA-SPIM is a new innovation that introduces new engineering design principles to biological microscopy for the first time. In the future, it is possible to push these principles much further. For example, the StarLith1900i water immersion lithography lens, with a staggering NA = 1.35 and field of view of 26 × 33 mm^2 (SBP = 594 gigavoxels), has been developed for patterning integrated circuits onto silicon wafers with 193 nm light [17]. Similarly, the large synoptic survey telescope (LSST) contains a record setting 3200 megapixel camera sensor [18]. While it is unlikely that these technologies are directly useful for biological microscopy, they set an example for how far the SBP limit and volumetric imaging barrier can be pushed, given appropriate engineering investments and scientific demand.

Conflict of Interest

The authors have no conflicts of interest to declare that are relevant to the content of this chapter.

References

1. Yueqian Z et al (2019) Systematic design of microscope objectives. Part I: system review and analysis. Adv Opt Technol 8:313–347
2. McConnell G et al (2016) A novel optical microscope for imaging large embryos and tissue volumes with sub-cellular resolution throughout. Elife 5:e18659
3. McConnell G (2020) Where next for the Mesolens? Wiley Analytical Science. Microscopy and analysis
4. Tang L et al (2024) Curved light sheet microscopy for centimeter-scale cleared tissues imaging. bioRxiv 2024.04.28.591483
5. Fan J et al (2019) Video-rate imaging of biological dynamics at centimetre scale and micrometre resolution. Nat Photonics 13: 809–816
6. Ota K et al (2020) Fast scanning high optical invariant two-photon microscopy for monitoring a large neural network activity with cellular resolution. bioRxiv 2020.07.14.201699
7. Yu C-H et al (2024) The Cousa objective: a long-working distance air objective for multiphoton imaging in vivo. Nat Methods 21:132–141
8. Yu C-H et al (2021) Diesel2p mesoscope with dual independent scan engines for flexible capture of dynamics in distributed neural circuitry. Nat Commun 12:6639
9. Sofroniew NJ et al (2016) A large field of view two-photon mesoscope with subcellular resolution for in vivo imaging. Elife 5:e14472
10. Voigt FF et al (2024) Reflective multi-immersion microscope objectives inspired by the Schmidt telescope. Nat Biotechnol 42:65–71
11. Shantanu S et al Microscope market size, share, growth & analysis | forecast – 2027. https://www.alliedmarketresearch.com/microscope-market-A06654
12. Glaser AK et al (2022) A hybrid open-top light-sheet microscope for versatile multi-scale imaging of cleared tissues. Nat Methods 19: 613–619
13. Dean K et al (2015) Deconvolution-free subcellular imaging with axially swept light sheet microscopy. Biophys J 108:2807–2815
14. Wu Y et al (2013) Spatially isotropic four-dimensional imaging with dual-view plane illumination microscopy. Nat Biotechnol 31(11): 1032–1038
15. Keller PJ et al (2010) Fast, high-contrast imaging of animal development with scanned light sheet-based structured-illumination microscopy. Nat Methods 7:637–642
16. Glaser AK et al (2023) Expansion-assisted selective plane illumination microscopy for nanoscale imaging of centimeter-scale tissues. Elife 12:RP91979
17. de Klerk J et al (2007) Performance of a 1.35NA ArF immersion lithography system for 40-nm applications. SPIE Proc 6520:742–755
18. Ivezić Ž et al (2019) LSST: from science drivers to reference design and anticipated data products. Astrophys J 873:111

Chapter 7

RAPID Autofocusing in Light-Sheet Microscopy: An Implementation Guide

Ludovico Silvestri

Abstract

In light-sheet microscopy, the excitation and detection paths are uncoupled and need to be aligned independently to ensure that the sheet of light lies in the focal plane of the detection objective. However, the presence of the specimen itself can break this alignment, frustrating the very principle of this technique and introducing significant blur. RAPID (Rapid Autofocus via Pupil-split Image phase Detection) has been reported as an effective method to correct sample-induced defocusing without introducing overhead optimization time. Here, I discuss some practical aspects on the implementation of this method, with the aim of providing a guide for the researchers interested in integrating this system within their light-sheet apparatus.

Key words Light-sheet microscopy, Autofocus, Clearing, Image processing, Phase detection

1 Introduction

Light-sheet fluorescence microscopy (LSFM) has been a game changer for the life sciences. Thanks to its high imaging speed and reduced photobleaching, this method enables high-resolution 3D analysis of intact biological specimens at scale [1–3]. The basic principle of LSFM is planar illumination: only a thin sheet of light is illuminated in the sample, located at the focal plane of the detection microscope objective (Fig. 1a). This optical scheme implies that excitation and detection light paths are spatially separated, and in most implementations, they reach (or exit) the sample from different objective lenses.[1] The distinct advantage of this scheme is that optical sectioning can be afforded using a fast wide-field detection scheme. The downside, however, is that excitation and detection paths are not guaranteed to be mutually aligned, as it happens in confocal or two-photon microscopies. Several studies [7–10], and

[1] Notable exceptions are represented by HILO [4], OPM [5], and SCAPE [6], where a single objective lens is used. Also in these cases, however, excitation and detection lights follow distinct optical paths.

Ludovico Silvestri (ed.), *Light Sheet Microscopy*, Neuromethods, vol. 229, https://doi.org/10.1007/978-1-0716-5100-1_7,

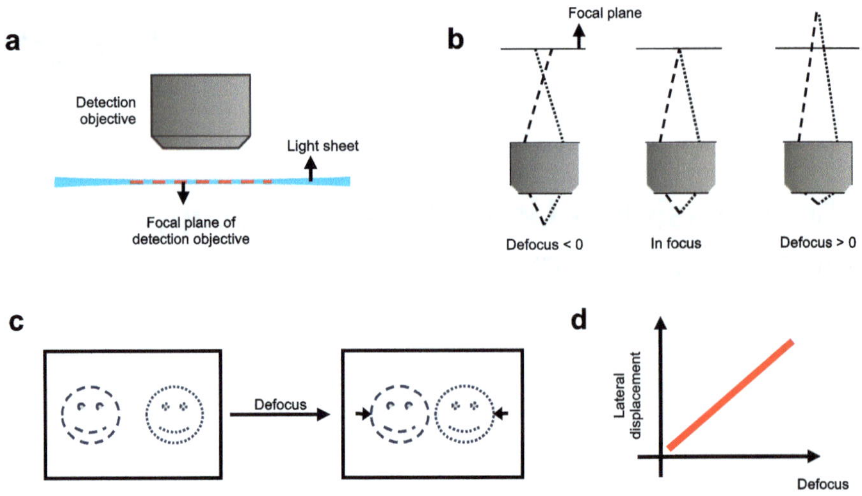

Fig. 1 Principles of phase detection autofocusing. Schematic representation of a light-sheet microscope (**a**), with the light sheet (light blue) and the focal plane of the detection objective (red dashed line) are ideally superimposed. Phase detection autofocusing stems from the observation that rays passing through different portions of the system pupil (here depicted with a dashed and dotted line) hit the focal plane at different lateral positions in presence of defocus (**b**). Thus, if two twin images are created using two different portions of the pupil (dashed and dotted smiling face), they will move laterally as the system goes out of focus (**c**). The linear relationship between the lateral displacement and the defocus enables using the former to univocally determine the latter (**d**)

daily experience of hundreds of LSFM users, reported that the sample can introduce a small but significant defocus in the microscope, de facto moving the sheet of light away from the detection focal plane (or vice versa).

This effect is particularly deleterious when imaging large samples at high resolution, where the depth of focus (DOF) of the detection optics is of a few microns and defocus varies a few tens of microns across the sample. Indeed, most LSFM users either image large samples at low resolution, where DOF is in the range of 10–100 microns [11–13], or select only a few areas where to perform high-resolution imaging, thereby manually adjusting the focus in these regions [14]. Other solutions reported in the literature include elongating the DOF by means of spherical aberration [15] or periodic optimization of the microscope parameters by maximization of image contrast [7, 8].

We recently reported RAPID (Rapid Autofocus via Pupil-split Image phase Detection), a real-time image-based autofocus that enables real-time correction of defocus in any wide-field microscope, including light-sheet ones [16]. RAPID is based on the well-known principle of phase detection, widely used in photography and telemetry (Fig. 1b): rays passing through different portions of the system pupil will intercept the image plane at different lateral positions when defocus is introduced. Thus, by physically separating these two beam groups and focusing them on two distinct regions of the same camera (or on two different cameras), it is possible to transform defocus into a mutual lateral displacement of these two images (Fig. 1c). The magnitude of this displacement is linearly proportional to defocus and can thus be used to restore focal position in real time (Fig. 1d) [16].

RAPID operates simultaneously to microscope operation, by introducing an additional optical system to the existing apparatus. Here, we describe in detail the optical and operational considerations needed to insert RAPID into any LSFM system, providing a guide for any researchers aiming at keeping images sharp throughout the entire specimens in an automated and *rapid* way.

2 Optical Design Considerations

The most direct way to insert RAPID into an existing light-sheet microscope is to place a beam splitting mirror in the infinity space between the detection objective and the tube lens (Fig. 2a). Since RAPID can operate at a slower frame rate compared to the main imaging camera [16], and thus will require a smaller amount of photons, my suggestion is to use a 90:10 (transmission:reflection) beam splitter to this aim. Potential workarounds to be used when this space is not accessible (e.g., in commercial systems) will be discussed at the end of the paragraph.

The RAPID system employs three lenses (L1, L2, L3). The first two are placed in 4f configuration to image the objective back focal plane (BFP) in a Fourier plane (F). The common focal plane between L1 and L2 is an intermediate image plane of the entire system (I1). The third lens (L3) acts as a tube lens, creating an image plane (I2) where the RAPID camera is placed. Denoting by f_{OB} the nominal focal length of the detection objective, and by f_{Li} the focal length of the i-th RAPID lens, the magnification of the system (at plane I2) will be:

$$M_{\text{RAPID}} = \frac{f_{L3} f_{L1}}{f_{L2} f_{OB}} \tag{1}$$

The magnification at the intermediate image plane I1 would instead be:

$$M_{\text{INT}} = \frac{f_{L1}}{f_{OB}} \tag{2}$$

It is also useful to consider the magnification of the Fourier plane F with respect to the objective BFP:

$$M_{\text{Fourier}} = \frac{f_{L2}}{f_{L1}} \tag{3}$$

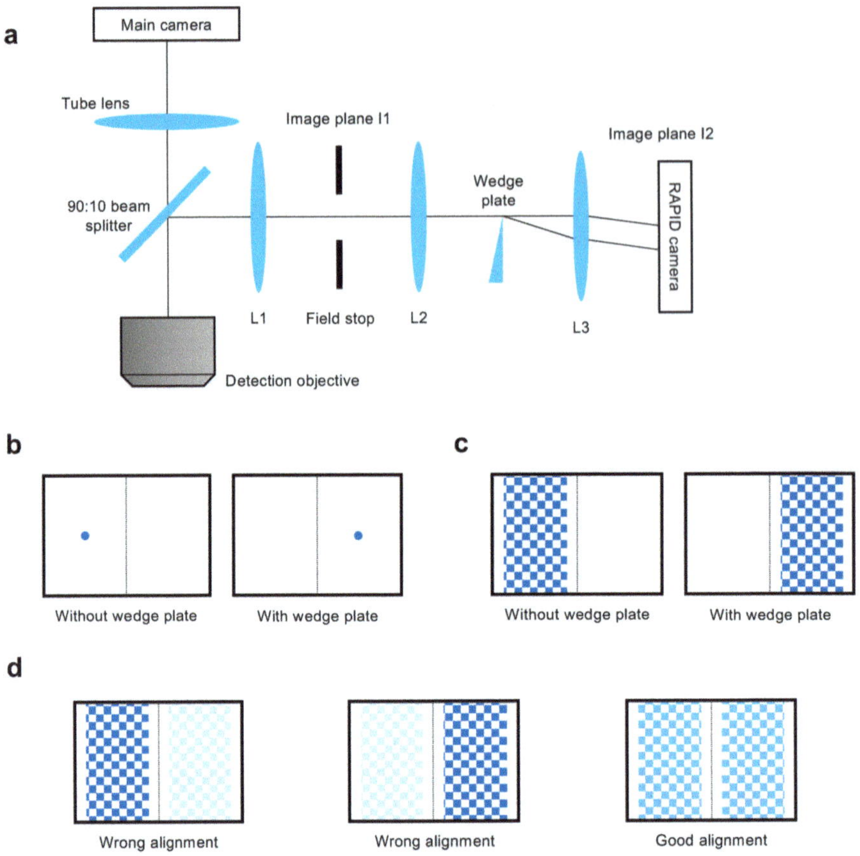

Fig. 2 RAPID implementation and alignment tips. Scheme of a standard RAPID implementation (**a**). Alignment of the system should be performed first with a laser beam (**b**). Then, the field stop should be inserted in the system and properly aligned (**c**). Finally, the right position of the wedge plate is found by equalizing the intensities of the two twin images on the RAPID camera (**d**)

Indeed, in the plane *F*, a wedge plate is placed. This plate is inserted to cover approximately half of the pupil image formed there, deviating about half of the rays and leaving the other half unhindered, thus creating the two pupil-split images required for RAPID operation.

As shown in [16], the lateral shift between the two images collected in plane I2 and the microscope defocus is given by:

$$\Delta x_{\pm} = \pm G\ NA\ M_{\mathrm{RAPID}} \Delta f \quad (4)$$

NA being the numerical aperture of the microscope objective, Δf the defocus, and *G* a factor of the order of unity. By denoting with *a* all the factors multiplying the defocus on the right side of Eq. 4, the distance between the two RAPID images is given by:

$$d = d_0 + a\Delta f \quad (5)$$

The mutual displacement at zero defocus d_0 depends on the angle β and refractive index n_p of the wedge plate placed in F [16]:

$$d_0 \approx \beta \left(n_p - 1\right) f_{L3} \tag{6}$$

A last useful equation to write down is the one concerning objective DOF:

$$\mathrm{DOF} = \frac{n_i \lambda}{NA^2} \tag{7}$$

λ being the wavelength in vacuum and n_i the refractive index of the imaging medium.

When choosing the different lenses to set up the RAPID system, several considerations should be taken into account:

1. *Sensitivity*

   The mutual displacement between the two RAPID images is measured using image registration algorithms. Assuming that image registration has an accuracy in the order of pixel size,[2] M_{RAPID} should be enough to guarantee that 1 pixel of displacement corresponds at least to the objective DOF. By combining Eqs. 1, 4, and 7, we find the following condition for the magnification:

$$M_{\mathrm{RAPID}} \geq \frac{NA\, \delta}{G\, n_i\, \lambda} \tag{8}$$

   δ being the pixel size of the RAPID camera. In general, we observe that the higher the RAPID magnification, the more sensitive the autofocus will be—i.e., it will detect smaller amounts of defocus.

2. *Robustness*

   Quantification of defocus is more robust the larger the field of view of the main apparatus ($\mathrm{FOV_{MAIN}}$) is collected by the RAPID camera. Indeed, if smaller FOVs are imaged, the risk of computing defocus using portions of the sample with scarce or null labeling increases. Since the field of view is the ratio of the camera size and the magnification, and the camera used for RAPID is unlikely to be much larger than the main camera of the microscope, the advice is to keep $\mathrm{M_{RAPID}}$ at the minimum level to satisfy Eq. 8.

3. *Throughput*

   The RAPID system operates with a fraction of light of the main apparatus. Thus, it is important to exploit it as much as possible. The wedge plate is placed to intercept about half of the pupil image in plane F (more details about its alignment will

[2] This is a convenient practical limit, although several studies demonstrated the possibility of sub-pixel accuracy in image registration [17, 18].

Table 1
Example of choice of the different components for a RAPID system

Variable	Value	Variable	Value
f_{OB}	18 mm	NA	0.6
f_{L1}	200 mm	β	5°
f_{L2}	150 mm	n_p	1.46
f_{L3}	100 mm	δ	6.9 μm[a]

[a]After 2 × 2 binning

be provided in the following paragraph). Since the edge of the plate will diffract part of the light—i.e., will deviate it in directions that will not be collected by the RAPID camera—it is important that the pupil image is significantly larger than 1 mm (assumed as an estimate of the wedge chamfer). As a rule of thumb, I suggest keeping $M_{\text{Fourier}} \gtrsim 1$, also considering that many objectives used in LSFM have large back apertures. As a reference, the radius ρ of the objective pupil is given by:

$$\rho = f_{OB}\ NA \tag{9}$$

As evident from these considerations, finding the best combination of focal lengths requires finding a trade-off between different requirements and the availability of specific optical components. In particular, wedge plates are available with an extremely limited choice of aperture angles. As a reference, I report in Table 1 the components used in the RAPID system integrated with the LSFM system present in my laboratory. The objective used is the XLPLN10XSVMP from Olympus-Evident (Japan) and the RAPID camera the Celera One CO5S-M from Alkeria (Italy).

A final consideration concerns the field stop to be placed in plane *I1*. This stop is needed to prevent the lateral portions of the two pupil-split images from being superimposed on the RAPID camera. The stop needs to be implemented only in the direction parallel to the displacement introduced by the wedge plate. In my lab, we have a custom slit machined from an anodized aluminum plate, whose size is slightly less than half the RAPID camera size (considering the magnification introduced by *L2* and *L3*). A variable slit, as produced by many optics manufacturers, can work equally well.

3 Alignment Remarks

The RAPID system is relatively easy to align for an experienced microscopist, as it requires putting in the right place three lenses to form a cascade of 4f systems. This can be easily achieved entering the system with a collimated laser beam. However, the presence of the wedge plate requires a few additional considerations.

Since the wedge plate will deviate part of the rays, it is convenient to align *L3* in such a way that it is equally displaced with respect to the unhindered and deviated light. To this aim, the position of the beam after L3 with and without the wedge plate should be observed, and it should be symmetrically displaced with respect to the original position of the beam (Fig. 2b). The same approach should be used for lateral alignment of the RAPID camera.

After aligning all the lenses and the camera using a laser beam, the field stop and the wedge plate should be put in place while the microscope is collecting images from bright and stable samples.

With the wedge plate out of the system, move the field stop in the direction perpendicular to the optical axis until the image fills almost half of the camera (Fig. 2c). Insert completely the wedge plate (so that it deviates *all* the light rays) and check that another image is formed on the other half of the camera (Fig. 2c). Move the field stop along the optical axis until its borders appear sharp on the camera. Check iteratively lateral and axial alignment.

The axial alignment of the wedge plate is, by empirical evidence, not sensitive to millimeter or sub-millimeter motion. The plate can be axially positioned in the putative focal plane common to both *L2* and *L3*. For lateral positioning, one should observe the image pairs produced on the RAPID camera and move the wedge plate until they are of equal intensity (Fig. 2d).

4 Computational Considerations

RAPID calculates the current defocus by inverting Eq. 5. Although the different parameters could be calculated from the system specifications, it is more convenient to determine them empirically by measuring the mutual displacement at various known defocuses. In this way, one would take into account potential biases in the computation of d by means of standard image registration procedures.

Whatever method for image registration is, in principle, usable to measure d. However, the software implementation of RAPID reported in [16] and available at https://github.com/lens-biophotonics/RAPID-AF uses multiple image filtering followed by 2D cross-correlation. The different filters (thresholding, difference of Gaussian, Canny filter) are used to enhance the contrast of the images, improving the cross-correlation results. The parameters of the filters, as well as the choice of the filter(s) to be used, depend on the optical settings (first of all M_{RAPID}) and on the features of the sample under investigation. For instance, in the presence of filamentous structures (e.g., blood vessels or neuronal processes), the Canny filter is very effective. On the other hand, with ball-like objects (e.g., cell nuclei), the difference of Gaussian performs at its best. A good rule of thumb is to always use multiple filters: the algorithm available on the GitHub link mentioned above will compare displacements computed with different pre-processing filters and discard them if they differ more than a user-defined threshold. Another good practice is to create and store different settings of the algorithmic parameters relative to the most common samples used in the LSFM apparatus.

Once the defocus has been calculated, the control software must send the right command for its correction. This usually occurs either by moving the detection objective with some motorized stage or by moving the sheet of light with a galvo mirror or other optical scanning system. This step can be quite tricky in commercial systems, as they might often come with closed-source control software. Possible workarounds include switching the control of the motorized stage from digital to analog and control the motion with an analog voltage generated with a standard DAQ board. Otherwise, one might consider inserting an additional stage in the microscope, which can be fully controlled with the RAPID software. In any case, the interface between the RAPID system and the instrument control is an important issue that must be solved case by case.

In general, the RAPID camera does not need to be run with a specific timing, since the autofocus system runs in parallel and independently from the main microscope. A remarkable exception to this rule is reported below.

5 Alternative Implementations

In many commercial systems, the researcher has no access to the infinity space between the detection objective and the tube lens. In these cases, the RAPID optical pathway cannot be connected to the main microscope as described above. An alternative configuration consists of extending the main optical system along the detection path by placing two relay lenses to form a 1:1 telescope. The 90:10 beam splitting mirror can then be positioned between the two lenses (Fig. 3a). Of course, this implementation has the disadvantage of requiring additional space, which might be problematic. However, the use of mirrors to fold the optical path may be considered.

Another variant that can be of interest is the use of two separate RAPID cameras to capture the two pupil-split images (Fig. 3b). This configuration is particularly useful to increase the FOV used for the autofocus, providing higher stability to system operation. In this case, a roof-top reflecting prism can be used in place of the wedge plate, to provide large spatial separation between the two

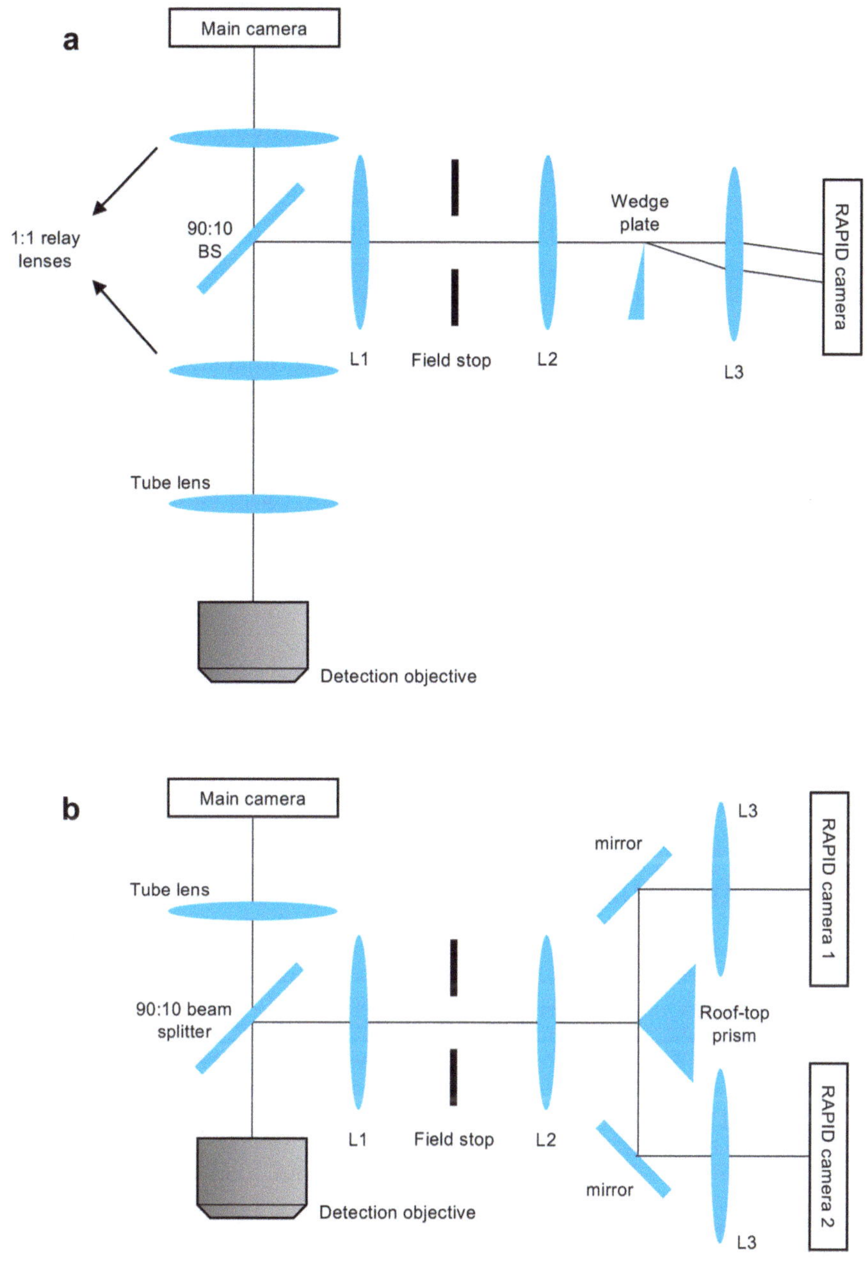

Fig. 3 Alternative implementations of RAPID. If the infinite space between the objective and the tube lens is not reachable, RAPID can be implemented using a 1:1 optical relay (**a**). To maximize the available field of view, 2 separate cameras can be used (**b**)

paths. From the computational perspective, it is important to run the two RAPID cameras with an external trigger, to ensure that the images that are further aligned with the spatial registration software correspond to the same portions of the specimen.

6 Conclusions

In this chapter, I provided a guide for the implementation of RAPID autofocus [16] on a generic light-sheet microscope. The choice of the different optical components and software parameters poses several trade-offs between sensitivity, robustness, and throughput and needs to be tailored to the single apparatus and application. Notably, the software parameters can be changed when changing the type of sample in a very flexible way, enabling use of the same apparatus for multiple projects.

The adoption of RAPID or of other stabilization systems [7, 8] in common laboratory practice would be an important step towards the routine generation of high-quality LSFM datasets, enabling this technology to reach the quality and reliability standards of older microscopy methods. In my view, this quantum leap is fundamental to release all the potential of light-sheet imaging for exploring the 3D structures of life.

Conflict of Interest

The authors have no conflicts of interest to declare that are relevant to the content of this chapter.

References

1. Keller PJ, Dodt H-U (2012) Light sheet microscopy of living or cleared specimens. Curr Opin Neurobiol 22:138–143. https://doi.org/10.1016/j.conb.2011.08.003
2. Hillman EMC, Voleti V, Li W, Yu H (2019) Light-sheet microscopy in neuroscience. Annu Rev Neurosci 42:295–313. https://doi.org/10.1146/annurev-neuro-070918-050357
3. Ueda HR, Dodt H-U, Osten P, Economo MN, Chandrashekar J, Keller PJ (2020) Whole-brain profiling of cells and circuits in mammals by tissue clearing and light-sheet microscopy. Neuron 106:369–387. https://doi.org/10.1016/j.neuron.2020.03.004
4. Tokunaga M, Imamoto N, Sakata-Sogawa K (2008) Highly inclined thin illumination enables clear single-molecule imaging in cells. Nat Methods 5:159–161. https://doi.org/10.1038/nmeth1171
5. Dunsby C (2008) Optically sectioned imaging by oblique plane microscopy. Opt Express OE 16:20306–20316. https://doi.org/10.1364/OE.16.020306
6. Bouchard MB, Voleti V, Mendes CS, Lacefield C, Grueber WB, Mann RS, Bruno RM, Hillman EMC (2015) Swept confocally-aligned planar excitation (SCAPE) microscopy for high-speed volumetric imaging of behaving organisms. Nat Photon 9:113–119. https://doi.org/10.1038/nphoton.2014.323
7. Tomer R, Ye L, Hsueh B, Deisseroth K (2014) Advanced CLARITY for rapid and high-resolution imaging of intact tissues. Nat Protoc 9:1682–1697. https://doi.org/10.1038/nprot.2014.123
8. Royer LA, Lemon WC, Chhetri RK, Wan Y, Coleman M, Myers EW, Keller PJ (2016) Adaptive light-sheet microscopy for long-term, high-resolution imaging in living organisms. Nat Biotechnol 34:1267–1278. https://doi.org/10.1038/nbt.3708
9. Power RM, Huisken J (2017) A guide to light-sheet fluorescence microscopy for multiscale imaging. Nat Methods 14:360–373. https://doi.org/10.1038/nmeth.4224
10. Ryan DP, Gould EA, Seedorf GJ, Masihzadeh O, Abman SH, Vijayaraghavan S, Macklin WB, Restrepo D, Shepherd DP (2017) Automatic and adaptive heterogeneous refractive index compensation for light-sheet

microscopy. Nat Commun 8:612. https://doi.org/10.1038/s41467-017-00514-7

11. Dodt H-U, Leischner U, Schierloh A, Jährling N, Mauch CP, Deininger K, Deussing JM, Eder M, Zieglgänsberger W, Becker K (2007) Ultramicroscopy: three-dimensional visualization of neuronal networks in the whole mouse brain. Nat Methods 4:331–336. https://doi.org/10.1038/nmeth1036
12. Renier N, Adams EL, Kirst C, Wu Z, Azevedo R, Kohl J, Autry AE, Kadiri L, Venkataraju KU, Zhou Y, Wang VX, Tang CY, Olsen O, Dulac C, Osten P, Tessier-Lavigne M (2016) Mapping of brain activity by automated volume analysis of immediate early genes. Cell 165:1789–1802. https://doi.org/10.1016/j.cell.2016.05.007
13. Stefaniuk M, Gualda EJ, Pawlowska M, Legutko D, Matryba P, Koza P, Konopka W, Owczarek D, Wawrzyniak M, Loza-Alvarez P, Kaczmarek L (2016) Light-sheet microscopy imaging of a whole cleared rat brain with Thy1-GFP transgene. Sci Rep 6:28209. https://doi.org/10.1038/srep28209
14. Glaser AK, Bishop KW, Barner LA, Susaki EA, Kubota SI, Gao G, Serafin RB, Balaram P, Turschak E, Nicovich PR, Lai H, Lucas LAG, Yi Y, Nichols EK, Huang H, Reder NP, Wilson JJ, Sivakumar R, Shamskhou E, Stoltzfus CR, Wei X, Hempton AK, Pende M, Murawala P, Dodt H-U, Imaizumi T, Shendure J, Beliveau BJ, Gerner MY, Xin L, Zhao H, True LD, Reid RC, Chandrashekar J, Ueda HR, Svoboda K, Liu JTC (2022) A hybrid open-top light-sheet microscope for versatile multi-scale imaging of cleared tissues. Nat Methods 19:613–619. https://doi.org/10.1038/s41592-022-01468-5
15. Tomer R, Lovett-Barron M, Kauvar I, Andalman A, Burns VM, Sankaran S, Grosenick L, Broxton M, Yang S, Deisseroth K (2015) SPED light sheet microscopy: fast mapping of biological system structure and function. Cell 163:1796–1806. https://doi.org/10.1016/j.cell.2015.11.061
16. Silvestri L, Müllenbroich MC, Costantini I, Di Giovanna AP, Mazzamuto G, Franceschini A, Kutra D, Kreshuk A, Checcucci C, Toresano LO, Frasconi P, Sacconi L, Pavone FS (2021) Universal autofocus for quantitative volumetric microscopy of whole mouse brains. Nat Methods 18:953–958. https://doi.org/10.1038/s41592-021-01208-1
17. Robinson D, Milanfar P (2004) Fundamental performance limits in image registration. IEEE Trans Image Proces 13:1185–1199. https://doi.org/10.1109/TIP.2004.832923
18. Pham TQ, Vliet LJ, van Schutte K (2005) Influence of signal-to-noise ratio and point spread function on limits of superresolution. In: Image processing: algorithms and systems IV. SPIE, pp 169–180

Part III

Practical Use of LSFM in Neuroscience

Chapter 8

Light-Sheet in a Multi-user Facility

Laura Batti and Stéphane Pagès

Abstract

Light-sheet microscopy enables high-resolution imaging of large, cleared tissues, such as whole brains, but requires careful preparation, mounting, and labeling strategies. In this chapter, we describe the challenges faced by a multi-user imaging facility and the solutions we found for providing access to light-sheet microscopy to users with different research goals. Running such a facility requires strong technical expertise, safety measures, and long-term maintenance yet offers unique advantages for advancing neuroscience and pathology through scalable 3D imaging.

Key words Light-sheet microscopy, Imaging facility, Sample mounting, COLM, MesoSPIM

1 Introduction

Light-sheet microscopy, applied to neuroscience, is a powerful technology providing insights into the structure and organization of biological circuits. In this chapter, we will discuss the challenges of running and managing a multi-user facility heavily based on innovative custom-built light-sheet microscopes and provide feedback from our experiences collected at the Advanced Microscopy Facility at the Wyss Center, Geneva.

In our core facility, we primarily focus on neuroscience applications and have integrated light-sheet systems tailored to fixed and cleared large samples. We process and image a variety of samples including whole rodent brains, entire rodent central nervous systems, and large human brain slabs. Our core facility aims to make our custom-made microscopes accessible to the scientific community, allowing researchers to address various scientific questions using the same systems. However, this democratizing activity poses several challenges that we will discuss in detail. These challenges are both technical—for example, multiple types of samples to prepare, multi-size samples imaging (from μm to cm)—and operational, such as running and managing a core facility based on non-commercial systems.

Ludovico Silvestri (ed.), *Light Sheet Microscopy*, Neuromethods, vol. 229, https://doi.org/10.1007/978-1-0716-5100-1_8,

2 Sample Preparation: A First Step for Good Light-Sheet Imaging

When integrating custom light-sheet microscopes in a multi-user facility, there are several points that need to be taken into consideration. Sample preparation for light-sheet microscopy goes beyond conventional protocols for histology and may involve processes such as clearing, sample embedding, and mounting in a specimen holder. These processes can vary depending on factors such as sample size and nature, microscope design, and the intended application [1]. The quality of light-sheet imaging is critically dependent on proper sample handling and preparation. In the case of imaging nervous system tissues, which are highly vascularized, an important upstream process that impacts imaging quality is the chemical fixation of the tissue. Suboptimal fixation can lead to an accumulation of strong autofluorescent signals from blood vessels, which can persist even after clearing protocols and result in nonspecific fluorescent signals throughout the tissue. Therefore, it is crucial to optimize and customize the sample preparation protocol for the specific application. This also applies to processes involving antibody labeling, where testing with appropriate negative controls is essential to verify signal specificity.

The number of protocols and methodologies for whole-organ clearing has expanded significantly in the past decade [2]. They span from hydrophobic methods (iDISCO+, for instance), hydrophilic methods such as CUBIC, and hydrogel-based methods, such as CLARITY. Some protocols are more suitable to preserve endogenous fluorescent signals, and others are more compatible with immunolabeling. As each specific application might require a tailored protocol, it is rather complex to find standard clearing methodologies to run routinely in a facility [2]. The implementation of clearing systems such as X-CLARITY© or SMARTBatch© is sometimes chosen by multi-user facilities to cover a specific spectrum of applications and to ensure a consistent outcome over time, as well as automation and batch processing.

Once the sample is optically cleared, before imaging, it is important to secure the sample in a stable manner to prevent movement during image acquisition. Depending on the microscope design, the sample may be held from the top or bottom, and in the latter case, sample stabilization within the cuvette becomes more challenging and requires precautions. This is particularly relevant when the sample density is lower than that of the imaging medium, causing it to float at the medium's surface. Embedding strategies in an agarose mold can be suitable for millimeter-sized samples, while larger samples may require different approaches such as gluing or anchoring one end of the sample.

Furthermore, the development of sample holders to meet specific needs is crucial. Therefore, guidance from facility experts and

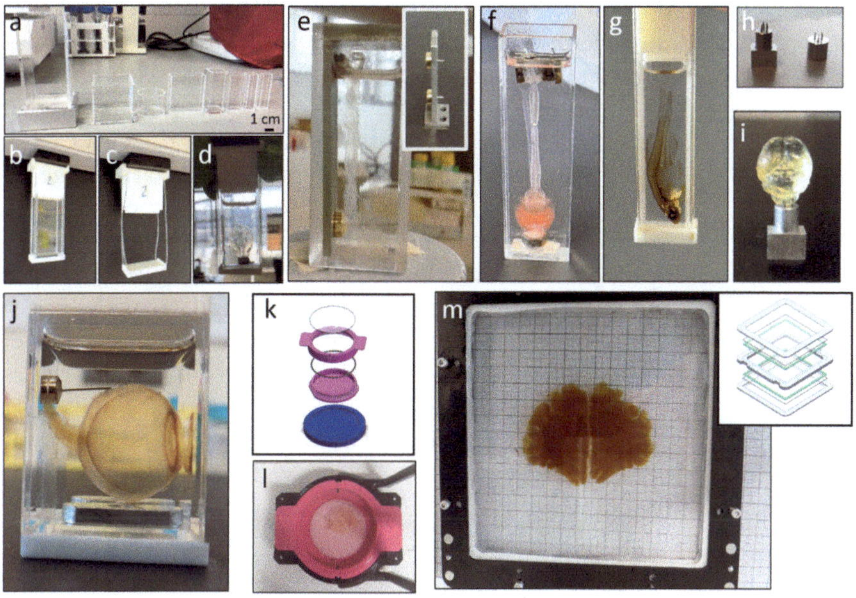

Fig. 1 Different sizes of custom quartz cuvettes for samples (**a**), cuvette assembled in a cuvette holder. (Design by Fabien Moreillon, HEPIA, Geneva) (**b**), and cuvette holder assembled with a magnet on top to match the microscope stage (**c**). Adult mouse brain cleared with X-CLARITY protocol glued at the bottom to a metal piece to keep the sample in place (**d**). Spinal cord cleared with X-CLARITY protocol pinched on the top and bottom by two needles system (insets top right) (**e**) (sample: N. Cho, EPFL). The whole mouse CNS cleared with X-CLARITY protocol, where the spinal cord and brain are glued to a magnet on top to a metal piece at the bottom, respectively. The magnet height can be adjusted by a metal bar outside the cuvette to obtain a straight positioning of the spinal cord (**f**) (sample: Q. Barraud, EPFL). Adult zebrafish assembled in a $1 \times 0.5 \times 5$ cm quartz cuvette (**g**) (sample: N. Regazzi, EPFL). Spiky sample holder to pinch stiff samples (**h**), mouse brain sample cleared with iDISCO+ mounted on a spiky holder (**i**), and human cleared eye glued at the bottom of the cuvette to a metal bar and stabilized at the top with a needle (**j**) (sample: M. Darche, Institut de La Vision, Paris). Exploded view of a sample holder for theta light-sheet microscope (**k**) and mouse brain mounted on a sample holder for theta light-sheet microscope (**l**) (by I. Gantar, Wyss Center). A human brain section mounted on a large specimen's holder for theta light-sheet microscope (**m**) exploded view of the assembled part on the inset (sample: C. Lamy and T. Jorda-Siquier, University of Geneva)

light-sheet specialists is essential to provide access to this technology for multiple laboratories working on different tissues and applications. Access to 3D printers and mechanical workshops is highly advantageous when working with both commercial and custom-built light-sheet microscopes.

In our facility, we have implemented different methods for handling samples based on their size, shape, and texture (hydrogel-based samples being more fragile than solvent-based cleared samples), as well as the microscope design (see Fig. 1). In all cases, the sample is physically isolated from the rest of the microscope's components by being enclosed in a quartz cuvette filled with the refractive index matching solution (RIMS), approach inspired by [4]. This mounting strategy offers the advantage of preventing the RIMS from coming into contact with the optics of the microscope, allowing the use of any type of media in small volumes (i.e., the volume of the cuvette). Without this isolation, RIMS based on organic solvents could potentially damage the optics and microscope parts. Additionally, the cuvette holders feature a magnet on top, enabling attachment to the stage of different microscopes. This allows users to obtain repeated images of the sample with different microscopes, such as those offering varying optical resolutions. This flexibility is crucial in an imaging facility like ours, where multiple users employ different clearing strategies. The samples are held in position by gluing them to metal pieces (typically for hydrogel-based samples) or by using needles or spiky holders for samples with a stiffer consistency. For very large samples, imaging chambers are used as cuvettes, and the samples can be fixed from the bottom or screwed onto the top of the sample holder.

Despite the valuable assistance provided by 3D printers and mechanical workshops, challenges remain when it comes to mounting unconventional samples. One such example is the imaging of expanded samples using expansion microscopy (ExM) [5]. This task proves particularly difficult for light-sheet microscopes with a T-shaped configuration, where the sample is held vertically. The expansion of the sample can exceed the limits of sample holders, and its low stiffness makes it challenging to stabilize in a vertical position without proper embedding. Open-top configurations are better suited for large samples of this nature, and gluing methods or embedding in agarose molds can aid in sample stabilization during imaging.

Another significant challenge in whole-organ clearing and sample preparation is related to labeling for multiple markers. Typically, light-sheet microscopes offer three to four lines of excitation (such as 405 nm, 488 nm, 561 nm, and 647 nm), allowing for imaging of a maximum of four different fluorophores. However, the 405-nm line often used to image DAPI-labelled nuclei is not very useful in thick samples due to the high diffraction of light at this wavelength, which is significant and prevents it from penetrating deep into tissues.

To surpass the limitation of the low number of excitation lines, various protocols involving cyclic labeling and stripping of probes have been recently developed [6, 7]. However, despite the availability of some commercially available kits, there are still no standard protocols to ensure multiplexing at the organ level. From the perspective of an imaging facility, we anticipate that this multiplexed approach, which is increasingly attractive to our users, presents significant challenges. For instance, the ability to reposition a sample in the exact same position and orientation between rounds of imaging is critical for subsequent realignment. An emerging technology that boosts the capabilities for multiplexing imaging is the integration of excitation lasers in the near-infrared (NIR) spectrum. This, other than adding a new acquisition channel to the scan, is extremely advantageous for imaging thick and turbid samples. Indeed, due to the physical properties of the wavelength, the light can penetrate deeper and with reduced scattering within the thick tissue. It has been demonstrated to enhance the information content of other fluorophores in live imaging experiments when combined with image restoration analysis [8].

Lastly, recent advances have addressed the long-standing challenge of performing RNA detection in cleared tissues. In particular, Kanatani et al. [9] developed a brain tissue–clearing method, TRISCO, which enables three-dimensional in situ hybridization on the whole brain, allowing visualization of specific mRNAs throughout the entire organ at cellular resolution. This development represents a significant step toward standardized RNA detection protocols compatible with whole-organ clearing and opens new frontiers for spatial transcriptomic analysis in large biological samples.

3 Light-Sheet Microscopy in a Multi-application Environment

In this second part of the chapter, we will focus on some of the solutions we have implemented to meet these various challenges in the context of a multi-application imaging facility. Indeed, integrating light-sheet microscopes into a platform for use by many researchers presents several additional requirements. As a matter of fact, during the last 10 years, a lot of different microscopes have been developed, each thought for a particular application. Among the specialized needs are microscopes for long-term in vivo imaging [10], clarified tissues, large samples imaged at moderate optical resolution, high axial resolution (lattice light-sheet microscope), and others. Because working in an imaging platform entails satisfying a wide range of users with specific needs, it is necessary to make some strategic decisions before acquiring these microscopes. One of the first features to consider is the optical resolution of imaging. Indeed, while some researchers may need to image very small structures like dendritic spines, others may want to image cells or networks of cells, or even larger organs. Some other researchers may also need to image large neuronal networks at subcellular resolution (e.g., mapping a single axonal fiber with axonal boutons and connecting distant brain areas).

To address these multi-user needs, there are two solutions: having different microscopes, each tailored for a particular resolution, or having a versatile microscope that can accommodate different resolutions. In the former case, the ensemble of microscopes should cover the entire range of resolutions. The main drawback of this approach is inherent to its cost and complexity in resource

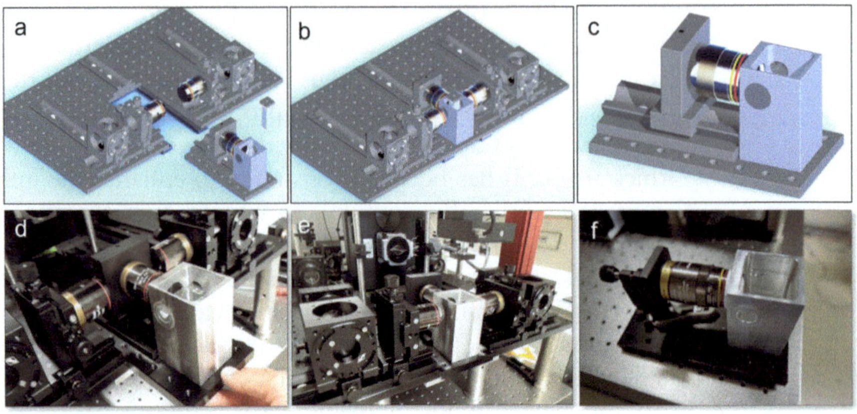

Fig. 2 3D rendering of the design of the modified version of COLM, during objective exchange (**a**) and during imaging (**b**). A detection objective and the imaging chamber are mounted in a stable configuration on the sliding breadboard to allow a fast exchange of samples and detection objectives. (**d–f**) The respective pictures of our actual system

management. In the latter case, the design of the microscope must be adapted. In our imaging facility, we have mixed both approaches. First, we have built a CLARITY Optimized Light-sheet Microscope (COLM) [4], which allows for high-resolution imaging at 0.6 μm of lateral resolution and 4 μm of axial resolution, and then, we have modified it by adding a new ability to quickly change detection objectives to modulate the optical resolution of the images. Modifications from the original prototype include a flip of the stage upside-down and the integration of a sliding platform where the detection objective and the imaging chamber are mounted (Fig. 2). We can now change detection objectives quickly (in approximately 1 min) and go rapidly from an in-plane optical resolution of 0.6 to 1.5 μm. With the original design, this simple operation required more than half a working day. Now each detection objective has its own platform that can be exchanged easily by lifting the sample out from the imaging chamber and removing the platform. A top magnetic adapter is used to attach the cuvette containing the sample to the xyz-theta stage. The cuvette is then immersed within the sample chamber filled with a matching RI medium. As discussed in a previous section, advantages of this approach are that different samples can be mounted without the need to remove the imaging chamber each time and without the need to empty and refill the chamber with RI medium. This addresses another significant challenge, which is the ability to image samples that have been clarified using different techniques, which require different RIMS. Additionally, the sliding platform configuration facilitates the imaging of the same sample and regions of interest using different detection objectives.

Another microscope covering a broad spectrum of optical resolutions is the mesoSPIM [11]. Thanks to a macro zoom system and the use of ETLs (electronic tunable lenses) which allow imaging of a large field of view (few square centimeters) while keeping an axial resolution around 3.5 μm, the mesoSPIM can image with a magnification that goes from 0.6× to 4×. The *in-plane* spatial resolution offered by this system ranges from 10 μm to 1.5 μm (at 0.6× and 4×, respectively). Note that the macro zoom has a relatively low NA at different magnifications (NA = 0.075 at 1× and 0.22 at 4×). A clear advantage of the mesoSPIM over other light-sheet microscope technologies lies in its speed of acquisition. An entire clarified mouse brain can be imaged in 7 min per channel at low magnification. COLM and mesoSPIM are two complementary microscopes. While COLM is used for high-resolution imaging (10× and 4×), mesoSPIM is used for fast prescreening of samples (0.6× to 4×). Indeed, having both within the same ecosystem allows a large sample screening pipeline where the samples are first imaged at a low resolution with mesoSPIM and then identified regions or cellular structures are selected for subsequent characterization at higher resolution using COLM. This enables a fast multi-resolution and data-efficient workflow. Another example of a versatile and multi-scale light-sheet microscope is the "hybrid" OTLS microscope [12]. In this case, the lateral resolution can be tuned from 0.5 to 10.7 μm in one single microscope where one objective is used for light-sheet illumination, and the other two objectives are used for orthogonal dual-objective (ODO) and non-orthogonal dual-objective (NODO).

As already mentioned, the variety of sizes, textures, and shapes of samples is another important consideration for an imaging facility. To image samples of several cubic centimeters, the microscope needs to have objectives for illumination and detection with long working distances (WD), when they have a T-shape configuration. For example, for COLM, a 2.9 cm working distance objective is used to create both light-sheets (XLFLuor 4X/340, Olympus), while the same objective is used for detection at 4×. An 8 mm WD objective is used for detection at 10× magnification (XLPlan N 10X/0.60, Olympus). Additionally, different cuvettes of different sizes are needed (see previous section). Note that this T-shape configuration, however, is not well suited for flat and thick samples like a human brain slice (ca. 15 × 10 × 0.7 cm), for which the motor stages need a large travel range and must move orthogonally to the detection path without interfering with the illumination objectives positioned obliquely with respect to the stage. To fulfill such a requirement, we have acquired an open-top light-sheet microscope, namely, the Clearscope from MBF originally developed by Migliori et al. [13]. This microscope is similar to the COLM, except for the configuration. It is an open-top theta light-sheet microscope. In this configuration, multiple tissue slabs can be accommodated in an array for high-throughput screening. In general, the open-top configuration also offers the possibility for integration of external devices that accommodate, for example, microfluidics, electrophysiology, and microdissection. This microscope is also complementary to the mesoSPIM and COLM, and having these three microscopes in the same environment allows imaging of various samples of different sizes and shapes at various optical resolutions.

Beyond all these technical considerations, one additional major challenge faced by an imaging platform is the sheer amount of data generated by the imaging process. Light-sheet microscopes, for instance, are known to generate very large datasets that can reach several terabytes per sample imaged. In an imaging facility environment, image storage issues must be anticipated very early and should involve IT (Information Technology) infrastructure that accommodates the processing and storage of the generated datasets. Many options exist to tackle this challenge, but two main solutions have emerged recently: an on-premise solution with a local data center and cloud-based solutions where data can be directly streamed to cloud storage. The advantages of this second option are the flexibility, scalability, and ability to benefit from high computing capabilities directly from the cloud. Sharing data is also an asset of using a cloud solution. Technical implementation complexity and costs are the main limitations of such a strategy. On the other hand, local storage is simpler and cheaper to implement but suffers from low scalability and the need for HPC resources.

Another dimension that needs to be considered is the fact that image analysis technologies are in constant evolution. Therefore, in a platform environment, it is important to keep up with these developments and implement them for the benefit of researchers. For example, in our platform, we use both commercial image analysis solutions and open-source software. It is crucial to maintain a balance between efficient and stable image analysis pipelines and new methods and technologies that are being implemented, which can be much less stable than commercial solutions. Striking a subtle balance between these two solutions is essential to benefit the platform users.

4 Light-Sheet Microscopy in a Multi-user Environment

In this part of the chapter, we will discuss the main advantages and disadvantages of having a custom light-sheet microscope integrated in a multi-user facility when compared to commercial light-sheet systems that are the most common solutions used in core facilities. The main comparison is made between commercially available solutions and microscopes already developed and designed by others and integrated afterwards within a core facility. Overall, the advantages of a custom-made microscope lie in its ability to offer tailored design, enhanced performance, and unique features.

The development of commercial instruments in the past 10 years has increased significantly, and many microscopes from a variety of manufacturers and suppliers are available today. Having a custom-built light-sheet microscope enables the introduction of novel and unique features in a system that can be further customized based on the requirement of the laboratory/facility. For instance, when we started developing our core facility, there was no solution available on the market to image the whole cleared rodent central nervous system (CNS) at subcellular resolution. We had the opportunity to collaborate with developers who designed light-sheet microscopes with unique features, and who were able to address those requirements. In our experience, having a core facility based on custom-made microscopes is also a source of inspiration for the scientific community, students, and engineers. We have organized courses for bioengineers, light-sheet workshops, and hands-on sessions for researchers. Together, those fostered the understanding of the principles of light-sheet microscopy and aided in building a community, attracting new collaborations, and spurring new innovations in the use of light-sheet microscopy.

5 Time, Expertise, and Human Resources

Integrating a custom light-sheet implies hardware and sometimes software development expertise within the team. At least one member of the team needs experience in setting up custom microscopes. Even though the cost of the microscope parts might fall in a lower budget than buying a commercial solution, the additional cost of personnel for the assembly, tests and maintenance needs to be taken into consideration. Timeline wise, if the system has already been designed by third parties, the most significant time and resources are used to order all parts, assemble them, and test the system. In our case, it required about 5–6 months for all parts to be ordered and received, a few days for the assembly, and a few weeks for testing. Before using the system for data production, we had a 3-week ramp-up phase where system and imaging workflows were defined and optimized.

Outside the building and testing, custom microscopes generally require involvement of specialized personnel for running the light-sheet activities: guidance on sample preparation, hardware calibration and maintenance, staff-assisted imaging activity, imaging quality assessment, and training to the use of the microscope. Therefore, a facility that includes custom light-sheet microscopes might require an exceptional workload of the employed personnel to take care of the activities mentioned above. From our experience, at least 50% full-time equivalent (FTE) is needed for running a maximum of two systems in a multi-user manner.

Because light-sheet imaging is generally associated with the production of large datasets, assistance and support of an IT department for the setting up of a light-sheet facility are essential to provide solutions for data handling, storage, and data sharing.

6 Reliability and Safety

When integrating a light-sheet microscope into a facility, it is important to anticipate if the use will be for biological data production or for R&D development in optics. Microscopes suited for biological applications need to ensure stability and generate reproducible data within long time frames. This might be in contrast with R&D purposes where the system is expected to undergo continuous modifications. Optimization and further customization of the systems can impact the stability of the system and jeopardize the reproducibility of long-term experiments. Commercial microscopes generally undergo extensive testing and quality control before they are released. This implies better reliability and more consistent performance than a custom microscope that might be subject to variability. With this in mind, we believe that it is very important to integrate quality checks and reproducibility tests also on custom-made microscopes and in the long run to keep monitoring the performance and stability of the systems.

In this context, worldwide community-driven initiatives were recently born aiming at improving reproducibility by developing standards in imaging and image processing and improving quality assessment and quality control such as point spread function monitoring and assessment of power illumination and stability [14–17]. Since custom-made microscopes are open and flexible to integrate new hardware and software features, it is also important to ensure the system stability, to have tight control of the upgrades, and to make sure that those will not irreversibly change the system output.

An option to ensure R&D progress and data production within the same ecosystem is to opt for a duplication of the equipment where the most expensive hardware elements are shared (i.e., camera and lasers) and the researchers can work on two separate microscopes.

Unlike commercial solutions, custom-made light-sheet microscopes might carry some hazardous features that need to be assessed, especially when the imaging includes the use of solvents and toxic chemicals. In our experience, following an audit from our onsite Health and Safety Officer, the following points were requested to be included before making the light-sheet microscopes available to the scientific community in a safe and controlled manner (Fig. 3):

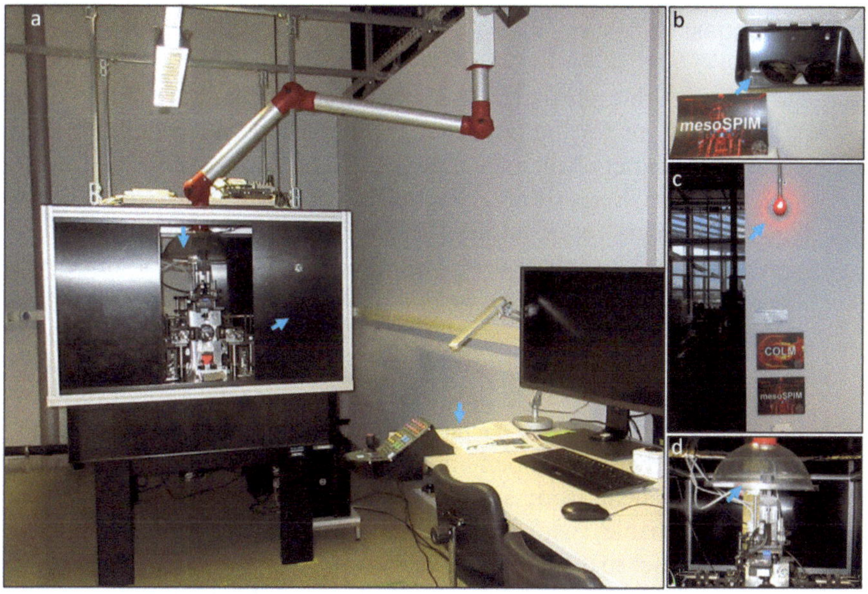

Fig. 3 The room hosting the mesoSPIM at Wyss Center, Geneva (**a**). Arrows indicate the safety-related features such as the aspiration hood above the microscope, the enclosure of the laser light path, and high desk to allow the good position of the user with respect to the light path. In the panels: protective goggles (**b**), flashing sign outside the light-sheet room (**c**), and aspiration hood (**d**)

- Controlled access to the microscopy area: the accessibility to the custom-built light-sheet needs to be highly controlled to ensure that all users are duly trained and updated with all work instructions. In our facilities, the systems are only available for trained researchers who can log in to the workstation and operate the microscope. Access to laboratory areas is restricted by badge recognition.
- Enclosure for the laser light path. When the light path of the microscope system is not physically enclosed, an interlock to laser or shutter can be implemented for major laser safety.
- Protective goggles for the alignment activities.
- Flashing sign outside the light-sheet room to indicate when the system is up and running.
- An aspiration hood at the level of the specimen's holder allows the use of hazardous chemicals such as DBE or THF. In our facility, each light-sheet has a separate aspirating hood on the top of the sample holder.
- The positioning of a desktop at a height preventing the laser path is at the same level of the user's face.

7 Unmet Challenges

Although there are many advancements in technologies and solutions for light-sheet microscopy and data handling, there are still unmet challenges that need to be addressed in the future.

Even though there is a common effort from software developers to create interfaces easily accessible to users from different backgrounds, user friendliness is still one of the crucial points distinguishing a commercial solution from a custom-made one. Regarding the assistance and maintenance support, in most cases, custom light-sheet microscopes rely on the activity and responsiveness of a single individual (hardware/software developer), making it difficult for facilities with low operational capabilities. On the other hand, open initiatives like the openSPIM and mesoSPIM that benefit from a large active community of users and developers enable an exhaustive assistance and technical support, including user manuals, online resources, and questions and answers, ensured by the community itself. This can be a source of inspiration for other custom-made microscopes.

In the long term, the lack of guarantee in the maintenance of custom light-sheet microscopes might bring system management complications. To obviate these potential issues, there are upcoming promising solutions proposing software and driver integration within one commercial solution. People might opt for this hybrid solution, where the custom-made hardware is driven by commercial software, to ensure more long-term stability of the microscope and better user friendliness.

Even if light-sheet microscopes can be quite versatile in terms of multi-resolution and size of sample that can be allocated, both commercial and custom light-sheet microscopes are suited either for live imaging or for cleared and fixed sample imaging. Those two macro applications can hardly coexist in a single microscope. Modular systems such as Flamingo from the Huisken lab at the University of Göttingen can provide versatility and offer a broader spectrum of applications within the same system. This can be the source of inspiration for new system development to support a variety of samples and biological applications (https://huiskenlab.com/flamingo/). Moreover, one of the most ambitious aspirations for a microscope user is to automatically combine low-resolution mesoscopic imaging and high-resolution imaging in specific regions of interest. This will be possible when microscopes will be implemented with intelligent feedback between control, acquisition, and data processing [18] and will eventually allow a fast large-scale screening of samples at high resolution without challenging the storage infrastructures.

Finally, one of the major challenges in light-sheet microscopy is the storage and sharing of large datasets. Despite advancements in technologies and solutions for large-scale databases [19], they still fall behind the rapid development of modern microscopy tools that generate large datasets at a high pace. Careful planning is necessary to define the infrastructure for data collection, database options, storage, backup, and sharing to ensure data safety and accessibility for different users.

Efficient strategies for managing large datasets involve cloud-based solutions and the deployment of large clusters that enable data to be written on the fly [20]. Cloud-based strategies offer scalability and accessibility, while large clusters provide high-performance computing capabilities. To facilitate data sharing, innovative solutions for data formatting have also emerged, specifically designed for the storage of large N-dimensional-type arrays in the cloud [21, 22]. These solutions address the challenges posed by data heterogeneity and fulfill the growing demand for community cooperation and interoperability in large dataset analysis, particularly in the context of AI-based analysis. Indeed, GPU is more and more required to run AI-based image analysis. This entails having access to complex resources involving modern and powerful graphics cards. Here again, cloud solutions are rapidly evolving and now offer very efficient image analysis pipelines. In fact, as mentioned above, beyond storage, it is also necessary to consider the steps and computing requirements involved in preprocessing images, such as stitching, alignment, and atlasing. In parallel, community-driven initiatives such as the Janelia Open Science Software Initiative and Allen Brain Initiative are promoting the sharing of open-source code and large data to enable reproducible image analyses across laboratories and to facilitate the widespread adoption of computational tools for large-scale imaging data. The larger the dataset, the more difficult these operations become, requiring computers that can handle the task. Data compression techniques can be part of the solution and can be advantageous for a platform environment, as they can speed up image processing and make it easier to exchange data with different groups of researchers using the platform [23, 24].

In conclusion, from the perspective of a multi-user imaging facility, we believe that offering multiple capabilities for whole-organ tissue preparation, multi-resolution light-sheet imaging, and data processing to a scientific community allows us to tackle neuroscience-related questions with innovative and unbiased approaches and to attract scientists coming from different fields. The versatility of the instrumentations and complementarity among the different microscopes can offer a multi-user interface. Integrating custom-built solutions in a multi-user facility ensures uniqueness and innovation; this comes with a luggage of key points that cannot be left out: a team with expertise in building and maintaining the instrumentation, the addition of quality protocols to ensure the reproducibility of experiments, integration of safety measures, and careful planning of data management.

Conflict of Interest

The authors have no conflicts of interest to declare that are relevant to the content of this chapter.

References

1. Reynaud EG, Peychl J, Huisken J, Tomancak P (2015) Guide to light-sheet microscopy for adventurous biologists. Nat Methods 12:30–34
2. Ueda HR et al (2020) Tissue clearing and its applications in neuroscience. Nat Rev Neurosci 21:61–79
3. Gantar I, Nguyen J, Karamanlis D, Ceto S, Osterop SF, Rajot D, El-Boustani S, Batti L (2025) Protocol for reproducible EZ clearing and labeling, including optimized steps and quantified fluorescence retention on mouse tissue. STAR Protoc 6(4):104105. https://doi.org/10.1016/j.xpro.2025.104105
4. Tomer R, Ye L, Hsueh B, Deisseroth K (2014) Advanced CLARITY for rapid and high-resolution imaging of intact tissues. Nat Protoc 9:1682–1697
5. Chen F, Tillberg PW, Boyden ES (2015) Expansion microscopy. Science 347:543–548
6. Wang X et al (2018) Three-dimensional intact-tissue sequencing of single-cell transcriptional states. Science 361:eaat5691
7. Saka SK et al (2019) Immuno-SABER enables highly multiplexed and amplified protein imaging in tissues. Nat Biotechnol 37:1080–1090
8. Gritti N, Power RM, Graves A, Huisken J (2024) Image restoration of degraded time-lapse microscopy data mediated by near-infrared imaging. Nat Methods. https://doi.org/10.1038/s41592-023-02127-z
9. Kanatani S, Kreutzmann JC, Li Y, West Z, Larsen LL, Nikou DV, Eidhof I, Walton A, Zhang S, Rodríguez-Kirby LR, Skytte JL, Salinas CG, Takamatsu K, Li X, Tanaka DH, Kaczynska D, Fukumoto K, Karamzadeh R, Xiang Y, Uesaka N, Tanabe T, Adner M, Hartman J, Miyakawa A, Sundström E, Castelo-Branco G, Roostalu U, Hecksher-Sørensen J, Uhlén P (2024) Whole-brain spatial transcriptional analysis at cellular resolution. Science 386 (6724):907–915. https://doi.org/10.1126/science.adn9947
10. Strnad P et al (2016) Inverted light-sheet microscope for imaging mouse pre-implantation development. Nat Methods 13:139–142
11. Voigt FF et al (2019) The mesoSPIM initiative: open-source light-sheet microscopes for imaging cleared tissue. Nat Methods 16:1105–1108
12. Glaser AK et al (2022) A hybrid open-top light-sheet microscope for versatile multi-scale imaging of cleared tissues. Nat Methods 19: 613–619
13. Migliori B et al (2018) Light sheet theta microscopy for rapid high-resolution imaging of large biological samples. BMC Biol. 16:57
14. Nelson G et al (2022) Monitoring the point spread function for quality control of confocal microscopes. https://www.protocols.io/view/monitoring-the-point-spread-function-for-quality-ccsuswew
15. Nelson G et al (2021) QUAREP-LiMi: a community-driven initiative to establish guidelines for quality assessment and reproducibility for instruments and images in light microscopy. J Microsc 284:56–73
16. Gaudreault N et al (2021) Illumination power and illumination stability v1. https://doi.org/10.17504/protocols.io.bzp8p5rw
17. Jambor HK (2023) A community-driven approach to enhancing the quality and interpretability of microscopy images. J Cell Sci 136 (24):jcs.261837. https://doi.org/10.1242/jcs.261837
18. Scherf N, Huisken J (2015) The smart and gentle microscope. Nat Biotechnol. 33:815–818
19. Allan C et al (2012) OMERO: flexible, model-driven data management for experimental biology. Nat Methods 9:245–253
20. Müllenbroich MC et al (2015) Comprehensive optical and data management infrastructure for high-throughput light-sheet microscopy of whole mouse brains. Neurophotonics 2: 041404
21. Moore J et al (2023) OME-Zarr: a cloud-optimized bioimaging file format with international community support. Bioinformatics 2023. https://doi.org/10.1101/2023.02.17.528834
22. Besson S et al (2019) Bringing open data to whole slide imaging. In: Reyes-Aldasoro CC, Janowczyk A, Veta M, Bankhead P,

Sirinukunwattana K (eds) Digital pathology, vol 11435. Springer, pp 3–10
23. Pomarico E et al (2022) Statistical distortion of supervised learning predictions in optical microscopy induced by image compression. Sci Rep 12:3464
24. Scholler J et al (2023) Efficient image analysis for large-scale next generation histopathology using pAPRica. Bioinformatics 2023. https://doi.org/10.1101/2023.01.27.525687

Chapter 9

Cell-Resolution Human Brain Reconstruction with Light-Sheet Fluorescence Microscopy

Irene Costantini, Luca Pesce, and Francesco Saverio Pavone

Abstract

The human brain presents an incredible structural complexity that enables the amazing richness of behavior. The organization of this organ spans multiple scales and requires tools that enable the reconstruction of centimeter-size volumetric blocks with cellular resolution. Here, we overview the methodologies that allow 3D human brain imaging using light-sheet fluorescence microscopy (LSFM). Specifically, we describe the SHORT method that enables the obtainment of optically transparent and macromolecule-permeable human brain tissues, in combination with tridimensional reconstruction using an inverted custom-made light sheet fluorescence microscope, providing the reader with all the preparation steps needed. We review the challenges related to the analysis of the big data generated by the acquisitions, providing guidelines and indications of solutions. Finally, we present an application on a human Broca's area as an example of the possible use of tissue clearing and LSFM for cellular mapping in 3D.

Key words Human brain, Tissue clearing, Immunostaining, Light-sheet fluorescence microscopy, SHORT, Broca's area

1 Introduction

Despite the substantial advantages that came with the introduction of automatic histology instrumentation, traditional protocols based on paraffin-embedding, tissue sectioning, and staining still suffer from significant drawbacks. Such limitations are inherent to the two-dimensional nature of the classic slide-based protocol including low sensitivity for sparse features, difficult assessment of dimensions, alteration of morphology, visual artifacts (such as differences in orientation or distribution), and destruction of the specimen [1]. All these problems hinder a thorough examination of the sample and consequently the advancement of structural analysis.

To overcome the limitations of bidimensional reconstruction, several methodologies were developed to perform 3D reconstruction of biological tissues using a combination of clearing and staining methods with high-resolution light-sheet fluorescence

Ludovico Silvestri (ed.), *Light Sheet Microscopy*, Neuromethods, vol. 229, https://doi.org/10.1007/978-1-0716-5100-1_9,

microscopy (LSFM) [2, 3]. The principle of tissue clearing relies on the homogenization of the refractive index inside and outside the sample, to avoid light scattering. To achieve it, various approaches were presented; the majority of which include a lipids removal step [4, 5]. The elimination of the membrane also facilitates the penetration of fluorescent probes deep inside the sample, allowing volumetric staining.

The endeavor of finding reliable tissue-clearing and staining methods comes primarily from neurobiologists, who are facing a daunting task: deciphering how the brain structure leads to function [6]. Indeed, there is a fundamental relationship between the structure and function of biological systems [7]. Therefore, a deeper understanding of the relationship between structure and function is necessary to characterize brain mechanisms. Moreover, any dysfunction occurring under pathological conditions may be explained in terms of structural alterations [8]. Consequently, structural analysis permits the identification and development of novel therapeutic strategies.

The brain is the most complex organ of the human body, with hundreds of different cell types organized into intricate networks that connect areas several centimeters apart from each other. Such an organization is extremely difficult to study with a bidimensional approach. In this context, the combination of tissue clearing and light-sheet microscopy has already allowed us to image large portions of human brains and to better understand the alteration caused by neurological diseases [6]. In recent years, numerous efforts have been undertaken to develop techniques for tissue-clearing in order to enhance optical imaging of human brain samples. Broadly speaking, the current clearing methods can be categorized into three groups: organic hydrophobic tissue clearing, hydrophilic tissue clearing, and hydrogel-based tissue clearing [4, 6]. For example, CLARITY-based clearing techniques have enabled successful imaging of various types of neurons in the human brain, such as cortical pyramidal neurons, interneurons [9], cerebellar Purkinje cells [10], and granule cells in healthy samples. Also, the compatibility of SWITCH-based methods was demonstrated to clear and stain numerous neuronal markers in thick human brain slabs, both healthy and affected by malformations of cortical development [11–13]. Other brain-clearing methods like iDISCO [14] and ScaleS [15] have been attempted on Alzheimer's disease samples and provided evidence of the association between microglia and amyloid plaques, as well as facilitating the detection of diffuse plaques. More recently, to tackle the density of human tissue, researchers have utilized small detergent (CHAPS) [16] and solvent molecules for the delipidation of large specimens [17]. Also, by elasticizing human tissue, probe delivery is improved when the tissue is stretched, and it also offers improved preservation over time, enabling multiple rounds of labeling

[18]. Finally, expansion microscopy has proven effective in enhancing the quality of images from thin, human tissue sections [19, 20], investigating human brains at the nanoscale level.

The mentioned studies demonstrate the feasibility of tissue clearing on human brain samples but also illustrate the limitations of this approach in humans. First, good transparency requires a significant extension of the time needed for the clearing, compared with that needed in rodents. Second, human brain tissue is highly autofluorescent due to the prolonged fixation time and the presence of retained blood inside the vessels and pigments like lipofuscin and neuromelanin. Third, it is very difficult to obtain homogeneous immunostaining of samples thicker than a few hundred micrometers; moreover, many antibodies failed to work on cleared tissue due to epitope masking. Finally, the huge amount of materials required for the staining makes the analysis extremely expensive, creating an empirical limitation and a replicability problem due to the variation of the products from batch to batch. As a consequence of all these problems, the application of 3D imaging is usually performed on small portions of the human brain, and/or on a single case study, limiting the statistic and the solidity of the finding.

Therefore, although the advancements made so far are really promising, routine cellular resolution 3D imaging of the entire human brain is still not possible and remains an important challenge of our time. In order to achieve it, it is essential to implement reliable clearing methods that enable scaling up the applicability to the entire human brain in a reasonable time to analyze multiple samples. Moreover, it will be necessary to substitute the use of classical antibodies generated using traditional hybridoma-based methods [21]. Recombinant monoclonal antibodies (mAbs) increase the reliability and reproducibility of staining while reducing the cost, making whole-human-brain staining affordable. Indeed, mAbs can be produced by cloning antibody cDNA or synthetic sequences into expression plasmids, and the production can be performed in mammalian cell culture. In addition, light-sheet fluorescence microscopes need to be implemented using optics with much longer working distances and a larger field of view, to permit the reconstruction of an entire sample. Another important aspect to consider is the need of developing new computational approaches that enable the management, storage, and analysis of large amounts of data (tens of terabytes) produced during the acquisitions. Finally, in order to compare the analysis obtained on different samples, it is important to align the data to a reference atlas, preferably MRI-based so as to have also the possibility of in-vivo correlation.

The combination of experimental protocols based on optical tissue clearing and advanced microscopy with dedicated software tools for data analysis addresses the issues of more traditional histology protocols, enabling 3D cellular resolution reconstruction of the human brain. The fast, high-throughput, and informative automated characterization of tissue architecture will eventually result in obtaining more precise anatomical information and pathological description of the samples.

In this regard, in the following sections, we overview the SHORT (SWITCH–H_2O_2–antigen Retrieval–TDE) methodology [22], a procedure based on standard histological treatments that combines the SWITCH immunohistochemistry technique [13] and the TDE clearing method [23] to perform routine clearing and labeling of human brain tissues. The transparent-labeled samples are imaged with a custom-made inverted LSFM that enables fast cellular-resolution 3D reconstruction of the tissues. The images are then post-processed with an efficient software pipeline that permits the analysis of the data. Finally, as an example of a possible application of the methodologies, a case study based on the analysis of a human Broca's area is presented.

2 Materials

2.1 SHORT Tissue Clearing

Human brains are fixed in 10% formalin for a minimum of 90 days. After PBS washing, the samples are cut into subsequent 500-μm-thick slices with vibratomes (such as Leica VT1000 S or Compressome VF-800 Microtome or custom-made apparatus). Each slab is incubated first in the SWITCH-off solution (50% phosphate-buffered saline (PBS) titrated to pH 3 using HCl, 25% 0.1 M HCl, 25% 0.1 M potassium hydrogen phthalate, and 4% glutaraldehyde) and after 24 h in the SWITCH-on solution (PBS pH 7.4, 1% glutaraldehyde). After three washes for 2 h each in PBS at room temperature (RT), the specimens are inactivated by overnight incubation with 4% glycine and 4% acetamide at 37 °C. The samples are then washed in PBS three times for 2 h at RT and incubated in a clearing solution (200 mM SDS, 10 mM lithium hydroxide, 40 mM boric acid) for 4 days at 55 °C. After clearing, the samples are washed three times in PBS and 0.1% Triton X-100 (PBST) at 37 °C for 24 h and treated with hydrogen peroxide (30% v/v) for 1 h at RT. After three washes of 2 h each in PBS at RT, antigen retrieval is performed using pre-heated Tris-EDTA buffer (10 mM Tris base (v/v), 1 mM EDTA solution (w/v), 0.05% Tween 20 (v/v), pH 9) for 10 min at 95 °C. After cooling down to RT, the sections are washed in DI water for 5 min each and then equilibrated with PBS for 1 h.

2.2 Immunostaining of Cleared Sample

Multicolor staining of the samples is performed by incubating the sample with primary antibodies 37 °C for 7 days in PBST at dilutions ranging from 1:100 to 1:200 (detailed list of the antibodies and concentration can be found in the SHORT paper [20]).

Following three washes in PBST, each lasting 30 min at 37 °C, the samples are incubated for 5 days at 37 °C with the secondary antibodies conjugated with different Alexa Fluor dyes with a dilution of 1:200, then washed three times for 1 h each at 37 °C. The final transparency is obtained by soaking the samples in increasing solutions of 20%, 40%, and 68% (vol/vol) of 2,2′-thiodiethanol (TDE) in PBS for 1 day each at RT with gentle shaking. Nuclei staining such as DAPI, propidium iodide, or Sytox Green (dilution of 1:100) can be added in the last TDE incubation step. For section imaging and storage, samples are mounted in a sandwich-like configuration between a 250-μm-thin quartz coverslip (for refractive index matching at 1.46) and a standard glass sample holder, with a 500-μm-thick steel spacer in between. Storage can be done indefinitely; fluorescence stability has been demonstrated in samples stored for 1 year.

2.3 LSFM Imaging

Imaging is performed with a custom-made inverted light-sheet fluorescence microscope. The setup is equipped with two identical objectives inclined at 90° relative to each other and inclined at 45° relative to the sample holder plane. They alternately played excitation and detection roles. The objectives are a pair of LaVision Biotec LVMI-Fluor 12x PLAN with 12× magnification, NA 0.53, WD 8.5–11 mm equipped with a correction collar for refractive index matching with the immersion solution. The optical resolution is 1.1 μm lateral and 3.3 μm axial, and the field of view is 1.1×1.1 mm^2. The sample is held in a custom sample holder inserted into a plastic tray filled with refractive index matching medium (glycerol 91%). A three-axes motorized stage system (two PI M-531.DDG and a PI L-310.2ASD for a motion range within $30 \times 30 \times 2.5$ cm^3 with submicrometric repeatability) positioned the sample and translated it along the horizontal direction during the imaging. The image velocity is 47 frames/s, corresponding to a volumetric rate of 0.5 cm^3/h. The microscope is equipped with four laser sources: Cobolt MLD 405 nm/100 mW, MLD 488 nm/60 mW, DPL 561 nm/100 mW, and MLD 638 nm/180 mW. The fluorescence is specifically filtered before being detected by an sCMOS camera (Hamamatsu Orca-Flash4.0 v3). The acquisition is performed by acquiring subsequent stacks along the x-axis; after each stack, the sample is shifted by 1 mm along the y-axis, and the next stack is acquired. Contiguous stacks have an overlap of 100 μm that allows the fusion of various stacks in post-processing to form the whole volume.

2.4 LSFM Data Analysis

The data obtained with a 3D reconstruction of an entire slice with LSFM need a post-processing process. The LSFM objectives are inclined at 45° with respect to the sample plane; therefore, an affine transform is necessary to compensate for the 45° rotation and the motion during the measurement. The affine transformation brings the acquired volume back to the sample's coordinate system and performs a spatial down-sampling to 3.3 μm isotropic resolution. To obtain the entire representation of the acquired volume, the contiguous stacks need to be fused together using a custom-made stitching software called ZetaStitcher (G. Mazzamuto, "ZetaStitcher: a software tool for high-resolution volumetric stitching" https://github.com/lens-biophotonics/ZetaStitcher). To store the information, data are compressed using a JPEG2000 lossy approach with a 1:20 compression ratio. Sharing of all data is performed using open-access repositories such as DANDI (www.dandiarchive.org) and EBRAINS (www.ebrains.eu).

3 Methods

3.1 Human Brain Clearing

To ensure efficient tissue clearing and multiplexed imaging, it is essential to achieve chemically stable tissue through the use of small, non-ionic crosslinkers with multiple functionalities. For the tissue-transformation process, the ideal molecule must meet two crucial criteria for processing most human specimens: (1) it should be capable of penetrating the entire tissue without the need of the perfusion, and (2) it should exhibit a high level of molecular cross-linking to enhance the durability and stability of the sample. Glutaraldehyde (GA) satisfies these conditions. Since its introduction by Sabatini et al. in 1962, GA has been the most commonly utilized bifunctional aldehyde fixative. It possesses two aldehyde groups linked by a flexible chain of three carbon atoms (Fig. 1a), which allows it to induce greater protein cross-linking compared to formaldehyde. Notwithstanding its comparatively small size, within a physiological solution (pH = 7.4), GA's capacity to establish cross-links is predominantly restricted to the outer layers of the samples, caused by its high ability to cross-link amino acid residues and slow diffusion into the slab [13]. This outcome arises from the rapid interaction of GA with the sample, leading to its depletion before achieving substantial penetration throughout the tissue. Consequently, the formation of gel is lacking in the central regions of the specimens. Likely, being a bifunctional aldehyde, GA can also undergo reactions with water and self-reactions, and the nature of these reactions is influenced by different factors, including pH. Indeed, in acid solutions (pH 3), the two aldehyde groups of GA react with water forming a cyclic hemiacetal (Fig. 1a). This predominant form in acid solution is less reactive and less capable of cross-linking proteins. Consequently, the kinetic reaction of GA can be controlled to achieve a uniform gel formation. To this aim, a low-pH buffer (termed SWITCH-off) was developed that, in addition to controlling GA reactions within the tissue, inhibits the fixation reaction through the protonation of endogenous biomolecules [24]. The described process corresponds to the SWITCH

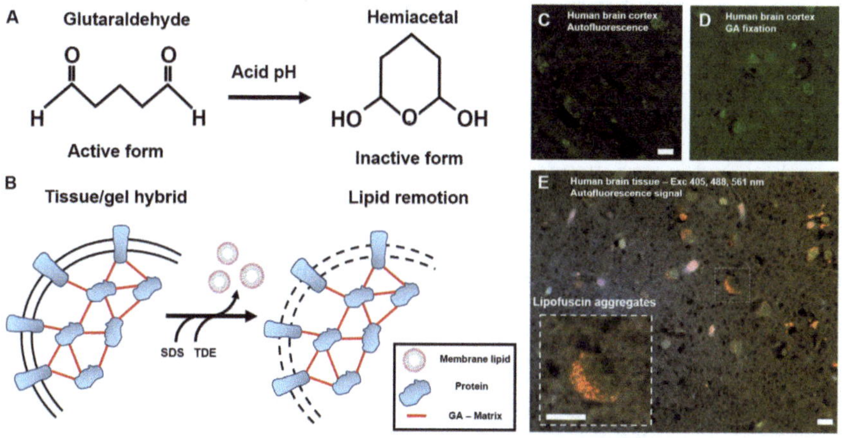

Fig. 1 SHORT enables the retention of cellular cytoarchitecture in 500 um human brain cortex samples. (**a**) The ability of GA to create crosslinks is strengthened under neutral pH conditions, while it is hindered at low pH, where it takes the form of a hemiacetal. By utilizing a low pH buffer, GA can uniformly spread throughout the sample, and subsequently, by adjusting the pH (pH 7.4), it forms a mesh-like gel structure throughout the entire specimen. (**b**) The accompanying illustration shows how GA can covalently bind within the sample. This process helps maintain the tissue's cytoarchitecture intact, and the removal of lipids can be achieved by employing a detergent and an index matching agent (TDE). A comparison of autofluorescence between paraffin (**c**) and paraffin + GA fixation (**d**) of human brain cortex samples. Paraffin + GA fixation reveals the presence of an undesired and spurious signal. Confocal images were captured using a confocal microscope, with an excitation light wavelength of 488 nm and a laser power of 1 mW. (**e**) The image depicts the cell body containing lipofuscin granules. Exc light at 405, 488, and 561 nm. Obj 100×/NA1.4, scale bar 10 μm. (The images were adapted from [22] and [12])

protocol [13], which revolves around leveraging the pH dependency of GA. This approach enables the dispersion of GA throughout a sample by temporarily inhibiting the cross-linking reaction using a low-pH buffer. Subsequently, by adjusting the sample's pH to a neutral value, the cross-linking activity of GA is reactivated (Fig. 1b). In SHORT [22], this kinetic OFF/ON reaction is optimized, in order to improve the architectural preservation and minimize the clearing time in 500 μm formalin-fixed human brain tissues. Details are specified in the next paragraphs.

The high structural preservation of GA has two important implications. Firstly, the use of an additional fixative can cause significant masking of important epitopes. Indeed, the efficient generation of intra- and intermolecular bridges of GA, which is crucial for tissue preservation, can lead to a loss of antibody immunoreactivity. Secondly, GA is known to induce a diffuse autofluorescence background, which can interfere with the acquisition of slabs (Fig. 1c, d). Both formaldehyde and glutaraldehyde fixation generate a broad emission spectrum that results in high background fluorescence. The formation of double bonds between the crosslinker and proteins increases autofluorescence (Fig. 1d), rendering the sample unsuitable for immunofluorescence assays. To mitigate this undesired effect, samples are incubated in an inactivation solution after SWITCH-ON, which terminates any ongoing fixation reactions within the tissue. However, there are additional factors that can contribute to nonspecific signals. Lipopigments and blood vessels, in addition to GA, can interfere during acquisition and produce false positives. Lipopigments, including lipofuscin (Fig. 1e), tend to accumulate in the aging human brain and cause endogenous autofluorescence that hampers the accurate identification of labeled neurons, leading to incorrect image interpretation. To address this issue, the sample must be treated with an autofluorescent quencher. Furthermore, SHORT exploits a combined approach between hydrogen peroxide and antigen retrieval. Hydrogen peroxide emerged as the best option for preserving tissue structure and effectively quenching autofluorescence. Additionally, SHORT combines the quenching effect with antigen retrieval to leverage its unmasking capability, enabling efficient staining for multiple biological features.

Such covalent linking of biomolecules based on GA fixation significantly decreases protein loss upon delipidation. Lipids, along with chromophores, can negatively impact imaging due to their contribution to refractive index mismatching, resulting in increased light scattering and light absorption. The removal of lipids enables easy matching of the refractive index, facilitating efficient light penetration and imaging at greater depths. In addition to the lipid removal procedure, matching the refractive index of delipidated tissue can enhance this process. In SHORT, we used a non-toxic agent, TDE, which enables sample dehydration and refractive index diminution (Fig. 1a). The absence of lipid barriers enhances the exogenous labeling of the tissue at deeper levels. Without lipid bilayers hindering the access of labeling agents, it becomes easier to introduce external probes for imaging with advanced optical techniques.

3.2 Whole-Mount Immunofluorescence Staining

Evaluating the suitability of antibodies for both neuronal and non-neuronal markers in human brain samples presents another hurdle. This is primarily due to the variability in post-mortem fixation conditions and the alterations that can occur to epitopes in such samples. In SHORT [22], the GA–tissue–gel, combined with the unmasking effect of antigen retrieval, enables the simultaneous detection of multiple markers in human brain slabs. The resulting tissue transformation process demonstrates effective staining for various biological features by incubating the antibodies at 37 °C. This temperature enhances the diffusion of the antibodies into the slabs, leading to homogeneous staining in the gray and white matter. With the utilization of this protocol, tissues can be successfully stained with a wide range of markers, encompassing both high-density proteins like neuronal nuclear antigen (NeuN) (Fig. 2a) and low-density proteins such as calretinin (CR). Also,

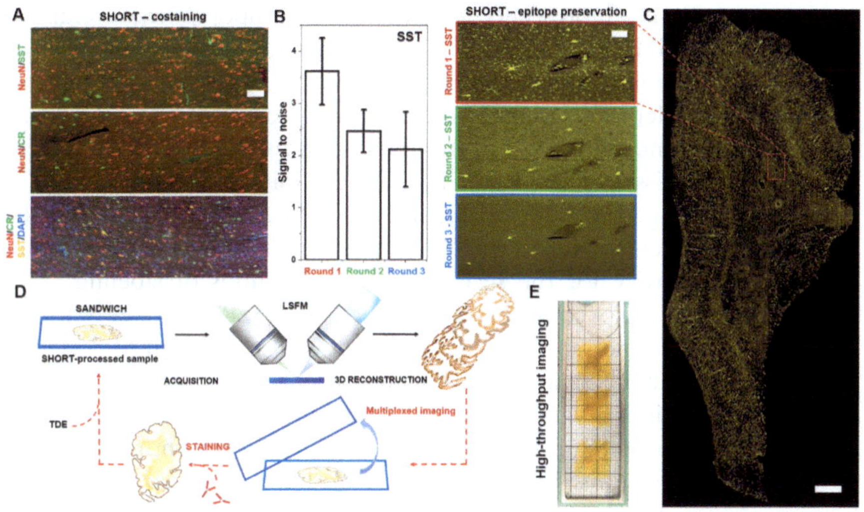

Fig. 2 SHORT allows multiplexed and high throughput imaging of the human brain cortex. (**a**) Different examples of optimized costaining in SHORT-processed slices. Maximum intensity projection (MIP) of 20 slices. Downscaled images, scale bar = 100 μm. (**b**) Three distinct rounds of immunostaining for probing the epitope preservation of the inhibitory marker SST. SNR was quantified for SST immunolabeling in round 1, round 2, and round 3 ($n = 30$ neurons). MIP of 70 slices. Downscaled images, scale bar = 100 μm. MIP of 70 slices. (The images were adapted from [22]). (**c**) Mesoscopic reconstruction of the processed human brain cortex slice labeled for somatostatin. Downscaled images, scale bar = 1000 μm. MIP of 70 slices. (**d**) Sketch showing the complete pipeline for processing and acquiring 500-μm-thick human brain slices. After clearing and staining, the tissue–gel hybrid is immersed in a refractive index matching (TDE/PBS) and enclosed in a sandwich holder to stabilize and preserve the specimens. Also, a multi-round of staining, stripping, and acquisition on SHORT-process specimens can be performed for proteome investigation. (**e**) Multiple SHORT-processed slices can be enclosed in the sandwich holder, to perform consecutive and high-throughput imaging using LSFM

SHORT enables co-staining protocols for the following combinations: vasoactive intestinal peptide (VIP) with somatostatin (SST) and parvalbumin (PV); NeuN with CR and SST; and glial fibrillary acidic protein (GFAP) with SST and vimentin (VIM) (some examples are reported in Fig. 2a). To prevent cross-recognition between the antibodies, different species can be selected for both the primary and secondary antibodies. Nevertheless, to explore the phenotyping analysis capabilities of SHORT, multi-round labeling is optimized (Fig. 2a). SHORT facilitates the comprehensive visualization of the proteome within large-scale human brain tissues [20, 22, 23]. Indeed, the tissue–gel hybrids produced through this method exhibit resilience to antibody stripping conditions, enabling the application of commercially available antibodies for multiple rounds of protein labeling and imaging, without the need for custom tagging. The removal of previously imaged antibodies is performed by incubating the sample in a clearing buffer at a temperature of 80 °C for a duration of 4 h. Also, this stripping procedure resulted in a reduction of approximately twofold in the autofluorescence signal emitted by lipopigments (e.g., lipofuscin). SHORT allows for the execution of three consecutive rounds of staining in 500-μm-thick slices of the human brain cortex [22], demonstrating effective antibody removal, preservation of neuronal morphology, and the ability to preserve the cytoarchitecture of the tissue (Fig. 2b, c). The tissue is initially co-stained with antibodies against VIP, SST, and PV. The antibodies are subsequently removed, and the tissue is restained, repeating the imaging process twice for the second and third rounds of labeling. The markers employed in these rounds are SST, CR, and NeuN for the second round, and SST, GFAP, and VIM for the third round, with SST acting as a fiducial marker to enable multiplexed profiling of the tissue. Various nuclear stains, including Sytox Green, DAPI, and propidium iodide, are utilized to fluorescently label human tissue sections. In recent works, we have compiled a collection of validated antibodies that underwent rigorous screening using advanced optical microscopy techniques on various human brain samples, demonstrating reliable and consistent staining [20, 22]. These antibodies offer a valuable resource for investigating diverse biological inquiries within the human cortex, providing a foundation for further research in this field.

3.3 3D Reconstruction with Inverted Light-Sheet Fluorescence Microscopy

The three-dimensional (3D) molecular and tissue characteristics of large human brain samples can be achieved by combining SHORT with light-sheet fluorescence microscopy (LSFM). The specific dimensions of human brain samples require a dedicated geometry compared to classical LSFM. Indeed, the sample has two dimensions that are significantly larger than the third dimension. Therefore, the specimens can be considered flat, but they require a custom-made system for the volumetric acquisition. An example is the dual-view inverted confocal light sheet fluorescence microscope, an apparatus specifically designed to image large samples measuring up to 30×30 cm^2, thanks to a spacious sample accommodation chamber [22, 25]. In a conventional LSFM setup, the sample is illuminated from one side with excitation light, while the emitted fluorescence is collected by an objective lens positioned at a right angle to the light sheet. The illumination axis is perpendicular to the detection axis. The sample is typically placed in a sample chamber or cuvette, with the light sheet passing through the sample in a perpendicular orientation. This setup poses challenges when acquiring samples placed on a glass-bottom dish or slide. To overcome this issue, the inverted LSFM setup was developed to exploit the dual views coupled with confocal-based detection. In this LSFM, the emission and excitation occur at 90° relative to each other and at 45° from the horizontal surface where the sample is placed. Both arms of the setup are used to simultaneously excite and acquire samples labeled with four different colors (two colors for each arm) (Fig. 2d). Additionally, by fusing the two orthogonal views, LSFM enables an isotropic resolution of 1 μm along the three optical axes (*x*, *y*, and *z*). Motorized optical filters are used to optimize the image contrast and avoid crosstalk between four fluorophores. A motorized stage moves the cleared specimens horizontally to acquire the volume scans. Tissue blocks from various brain regions are prepared by cutting them into slices up to 500 μm thick using a vibratome. These slices are then treated with SHORT and labeled using indirect immunofluorescence. The cleared samples are mounted in a custom-made sealed sample holder, called "sandwich," which permits imaging and long-term storage (Fig. 2e). The sandwich holder consists of a cover glass, a spacer with a thickness of 500 μm, and fused silica glass. The cleared specimen is fixed between the cover glass and fused silica glass using a special glue. Such a strategy allows soaking the sample in the TDE/PBS solution supplemented with a nuclear dye, and performing high-throughput imaging of several labeled human slices. Using the sandwich apparatus, we observed the preservation of fluorescence signals of SHORT-processed slices stored in the sandwich for 1 year. The tissue dehydration caused by the TDE solution had the beneficial effect of slightly reducing the specimen volume and preserving the tissue, which is advantageous for routine acquisition and long-term storage of labeled specimens. The LSFM system has a high-throughput capability and provided an isotropic resolution of $3.3 \times 3.3 \times 3.3\ \mu m^3$. This resolution enabled the visualization of single neuronal subtypes and the morphological features of their individual subpopulations, facilitating the 3D reconstruction of human cortical cytoarchitecture.

3.4 Challenges of Data Analysis and Management

The advancement of microscopy techniques has created a new problem usually constrained to other scientific fields: the handling of big datasets. The high-speed, high-resolution imaging of volumetric specimens produces a huge amount of data that need to be analyzed and managed. The need for new computational approaches is becoming more compelling every day. For example, the 3D reconstruction of a single human brain block of 1 cm^3 led to thousands of individual 3D subvolumes (called "stacks" or "tiles") and tens of terabytes of raw data. Indeed, the typical optical resolution used to perform cell counting or neuronal tracing is in the micrometer scale. To achieve such resolution, high-magnification objectives (5×, 10×, or 20×) are used. These objectives have a limited field of view that cannot be used to acquire the whole sample. Therefore, to perform the 3D reconstruction of large specimens, it is necessary to perform optical scanning of subsequent adjacent stacks that cover the whole volume. The acquisition is performed by either moving the sample or the optics, but it always requires overlap between adjacent regions to enable a post-processing stitching step of the tiles to obtain the final 3D reconstruction.

Once the 3D volume is acquired, the data need to be efficiently stored, analyzed, visualized, and shared. Each of these steps creates a different challenge that needs to be faced with different tools that are the focus of ongoing efforts in the field.

Concerning data management, various discussions have been raised about the need for standardized formats designed to facilitate access to the metadata and rapid compression and decompression of the images. An example is the BIDS standard [26, 27] that prescribes a format that helps organize neuroimaging data in a file system that simplifies communication between neuroscientists through consistent naming and location of file data. Nevertheless, a reduction in storage requirements and a speed-up of computational processing of image data are fundamental due to the large quantity of the data produced (tens of frames per second). To this aim, there is an ongoing discussion about which data need to be stored: the raw datasets acquired by the system or the post-processed images (compressed or not).

Another important challenge is the transformation of the raw images acquired by the microscope into a spatially coherent, high-resolution image volume that is suitable for data visualization and analysis. The individual stacks need to be precisely aligned and fused. This is even more complicated if complementary views or various channels were recorded by the microscope. An example is the ZetaStitcher tool (G. Mazzamuto, "ZetaStitcher: a software tool for high-resolution volumetric stitching" https://github.com/lens-biophotonics/ZetaStitcher), a software that permits to fuse the terabyte-scale image stacks acquired by the microscope using the overlapping regions of contiguous tiles obtaining the final 3D representation of the whole sample. However, unlike mouse brain reconstructions, human brain imaging is performed on thick tissue slabs obtained from serial sectioning of the tissue. This kind of acquisition creates another challenge: the alignment of the various sections with one other. This could be done using the internal landmarks of the image as a reference or by aligning the image to the reconstruction obtained from other modalities (in the next paragraph, this aspect is illustrated in detail).

Once these post-processing steps are complete, the high-quality image volume reconstructed needs to be analyzed to extract quantitative information. To this aim, several automatic processes, both supervised and unsupervised, have been proposed. Machine learning techniques based on neuronal networks require huge amounts of manual annotation but allow obtaining 3D neuronal segmentation at a pixel level [11]. On the contrary, unsupervised methods do not require labeled data and are more flexible in terms of applicability to different subjects and analysis of various markers [28] but are less precise. An intermediate tool is presented by the iLastik toolkit [29], a semi-supervised approach for image segmentation that adopted an interactive image classification and segmentation method where the user needs to label a small subset of the image but is able to visualize the results in almost real-time to refine the analysis.

Once quantitative information is obtained, a big challenge of human brain analysis is the comparison between different subjects. Indeed, contrary to mouse brain research, there are few human brain datasets available that can be used as referenced atlas. An attempt to do so is given by the EBRAINS platform created during the European Horizon 2020 Human Brain Project. The platform aims at collecting multilevel information into the same reference, providing a place where to integrate anatomical, connectivity, and functional data. However, the mapping can be performed on a single brain, the BigBrain [30], used as a reference and not on a common anatomical atlas. Another approach is provided by the possibility of using MRI imaging of the same subject as a reference to map the analysis performed on the optical reconstruction. This multimodal approach was used in a recent study to perform the cell census of a human Broca's area and is described in detail in the next section.

3.5 Application to Human Broca's Area

A large multi-institutional collaborative effort across the Massachusetts General Hospital, Boston University, the European Laboratory for Non-Linear Spectroscopy, and the Icahn School of Medicine at Mount Sinai has created the first cell census of different neuronal types of a human Broca's area within an MRI reference system. The study, published in Science Advances [25], uses a multimodal imaging infrastructure to integrate cellular anatomical information within a whole-brain reference space. Contrary to mouse brain reconstructions, which can be now achieved with commercial LSFMs, no current imaging technology can directly visualize microscopic features, such as specific cellular bodies, in the whole human brain. The dimension of the human brain requires processing the sample by cutting it in small portions to perform 3D reconstruction with custom-made LSFM apparatus. The combination of tissue slicing, clearing, labeling, and mounting produces tissue distortions that prevent an easy recognition of the laminal structure and cytoarchitectonic boundaries of the brain.

To compensate for these problems, a combination of various techniques was employed to bridge the gap between the macro- and the micro-scale. In particular, an intermediate technique, the optical coherence tomography (OCT), was used to register the volumetric 3D acquisitions obtained with LSFM to the MRI reconstruction of the whole hemisphere. 7T MRI was performed on the whole sample, then the Broca's area was dissected and analyzed with OCT [31]. This optical imaging technique permits obtaining anatomical information by measuring the backscattered photons from the sample and thus does not require the staining of the sample. Imaging was performed with 5-μm isotropic resolution in

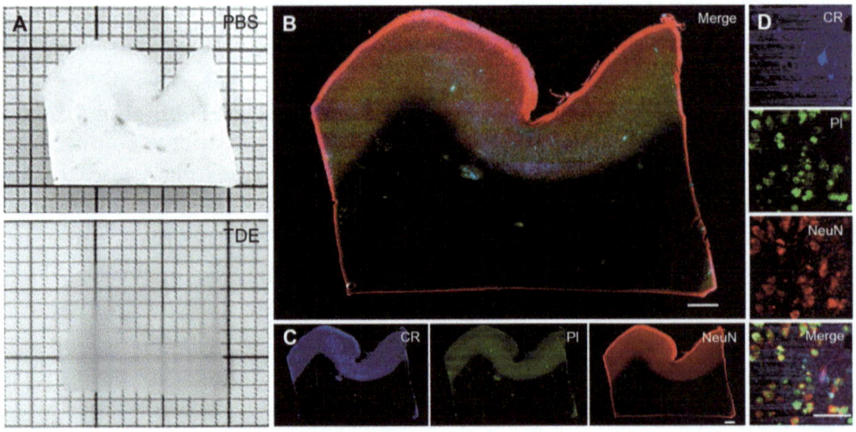

Fig. 3 3D imaging with LSFM. (**a**) Photos of a 500-μm-thick slice before and after SHORT clearing. (**b**) A representative slice of a middle plane (~200 μm depth) of the 500-μm-thick slice in **a**, stained with CR, PI, and NeuN (respectively in blue, green, and red). Scale bar = 1 mm. (**c**) Single channels of the slice in **b**. Scale bar = 1 mm. (**d**) High-resolution insets showing the different markers used in the study: calretinin (CR), propidium iodide (PI), and neuronal nuclear antigen (NeuN). Scale bar = 50 μm. (Figure adjusted from [25])

combination with a serial section of the tissue every 500 μm of depth, allowing the analysis of the centimeter-size block. The imaging is performed prior to slicing; thus, no structural alterations are introduced, permitting an accurate alignment of the sections and the registration to the MRI-based coordinate system. After the OCT acquisition, the slices are collected and treated for LSFM imaging. For this study, the 500-μm-thick slices were cleared with the tissue transformation technique SHORT and co-stained with three different labels (Fig. 3). An exogenous dye, propidium iodide (PI), to detect all nuclei was used in combination with an anti-neuronal nuclear antigen (NeuN) antibody to stain all neurons and an anti-calretinin (CR) antibody to identify a subpopulation of γ-aminobutyric acidergic (GABAergic) interneurons.

The inverted LSFM used in the study permits performing 3D reconstruction of the sample with an acquisition speed of 0.5 cm^3/h and 3.6 μm isotropic resolution (after post-processing). The analysis of the LSFM 3D reconstruction was performed with a stereological approach using an optical fractionator probe [32] adapted to analyze cleared materials. The counting performed on 3D reconstruction allows for visualization of the neurons in their entirety, leading to a precise quantification since over- or under-estimation biases diminished drastically. The total volume analyzed was $1.5 \times 1.3 \times 0.75$ cm^3 obtaining a counting of 22,461 CR+ neurons in layer 3 across the slabs, 85,321 NeuN+ neurons in layer 3, and 62,584 NeuN+ in layer 5, with corresponding densities of 2707, 8232, and 6274 neurons/mm^3, respectively (the counting obtained for each slice are available in the paper).

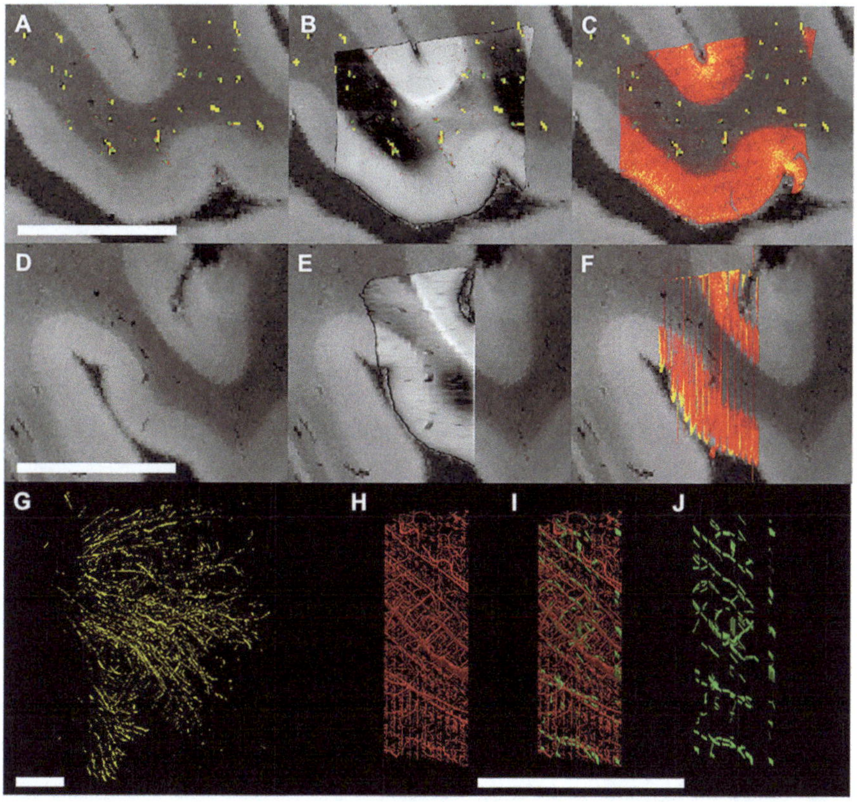

Fig. 4 MRI/OCT/LSFM alignment based on vasculature segmentation. (**a**–**c**) Vessel segmentation in the MRI, OCT, and LSFM volumes. (**d**–**f**) MRI, OCT, and LSFM reconstructed volumes across the slice direction (horizontally). (**g**, **h**, **j**) 3D reconstruction of the segmented vessels in each modality (MRI, OCT, LSFM). (**i**) Overlay of the OCT and LSFM segmentation after the nonlinear co-registration. Scale bars = 1 cm. (Figure from [25])

To refer the counted cell to their location in the MRI space, a nonlinear registration method was developed based on the segmentation of blood vessels. Indeed, vessels are visible in all imaging modalities and can be used as common features to align the different modalities with each other. In particular, blood vessels with a diameter larger than 150 μm were selected and manually segmented (*see* Fig. 4).

OCT measurements did not present distortion and are on the same scale as LSFM, creating an intermediate reference between MRI and LSFM. Hence, a first registration between MRI and OCT and between OCT and LSFM was computed, then the final MRI/LSFM alignment was performed using the information obtained from the other alignments. Indeed, all deformations are fully invertible, allowing to warp any modality into the space of any other modality. The final result is the placing of the cell counted with stereology on LSFM images within a cortical coordinate system created on the MRI reconstruction of the whole hemisphere as shown in Fig. 5.

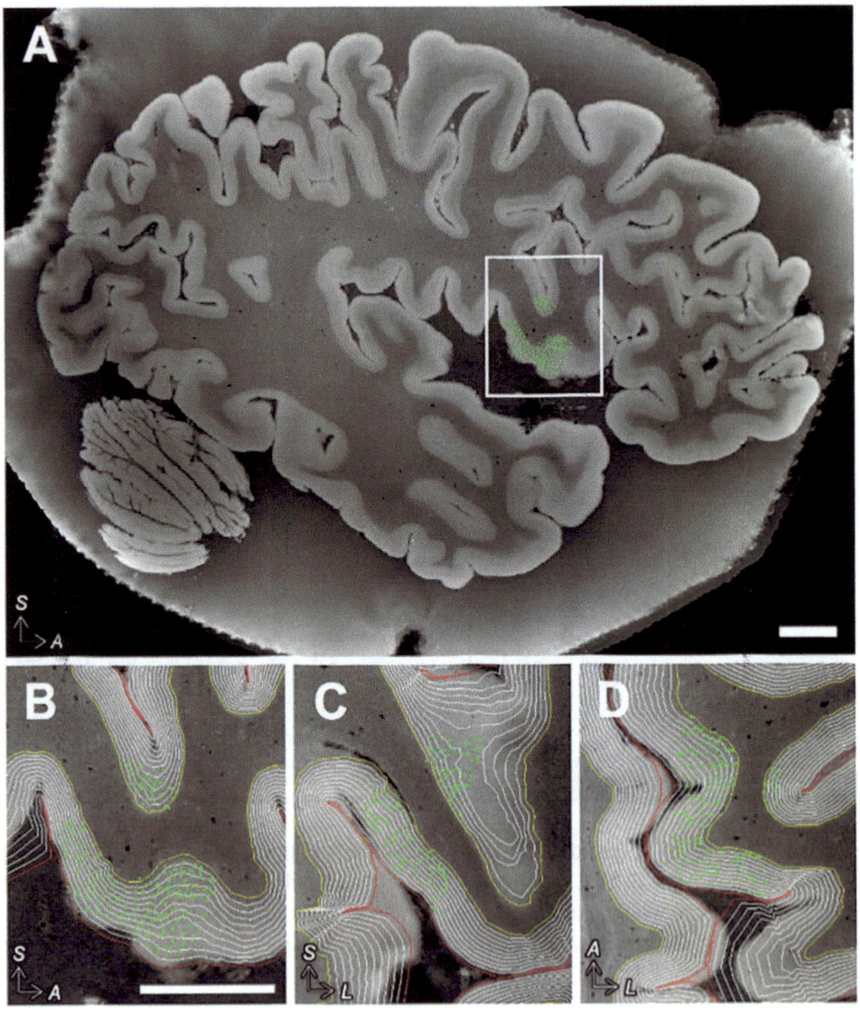

Fig. 5 Stereological counting sites in MRI space. (**a**) Mapping of the stereological counting sites (green) for the NeuN markers to the MRI space. (**b**–**d**) Projection of the counting sites in the sagittal, coronal, and axial plane. Scale bars = 1 cm. (Figure from [25])

3.6 Conclusions and Future Prospective

In the past decade, the clearing method field has seen an exponential spread. The innovations presented in tissue preparation combined with volumetric imaging have permitted the analysis of the brain cytoarchitecture in 3D with unprecedent resolution. The integration of knowledge in physics, chemistry, and tissue engineering has dramatically improved the development, permitting the creation of more reliable protocols compared to only a decade ago. Nevertheless, limitations and barriers remain to be overcome, in particular for human brain treatment. Homogeneous staining of the human brain in its integrity is still not possible due to the limitation of antibody penetration depth inside the tissue.

Moreover, the transparency achieved so far by the different methodologies is still not good enough to perform the reconstruction of centimeter-size volumetric blocks. The brain still needs to be cut into serial sections to be processed, thus requiring solutions to perform the alignment of sections to each other and with a common reference atlas. In the near future, we foresee the development of new clearing approaches to increment the size of the processed block to lower the distortions due to tissue cut, as well as the use of different staining methods to obtain more stable and deep labeling of the tissue. Nanobodies could be a solution to problems related to antibodies. The small dimension will permit an easy diffusion into the sample, and the known DNA sequence necessary to perform their production will permit to reduce the variability of the staining.

In parallel to tissue preparation, optics evolvement will also be necessary. Objectives with higher working distance and numerical aperture need to be developed to perform imaging of bigger blocks of tissue with micrometer resolution. Nevertheless, in order to achieve submicron resolution of large volumes, LSFM will need to be coupled with expansion microscopy. Up to now, expansion microscopy is employed to perform super-resolutions studies; however, for very dense tissue, such as the human brain's white matter, it is a possible solution to be exploited. An emerging possibility in scientific research is the exploration of post-expansion staining techniques for such dense specimens. This intriguing approach holds the potential to exploit the decrowding effects induced by the expansion process, leading to improved accessibility and density of protein staining within densely packed multicomponent structures. In particular, the connectome field will very much gain from the possibility of reconstructing centimeter-size portions of the human brain with single fiber resolution.

All the neuroscience fields will gain from the ability to investigate the 3D relationships of single components of the tissue. Still, the spread of these technologies will have their major implications for clinical applications. Indeed, the diagnostic potential of performing three-dimensional reconstructions of the brain will permit accurate histopathological investigations of samples. In conclusion, this work hopefully has provided guidance to encourage researchers to enter the field of optical human brain reconstruction, demonstrating the potential of this analysis.

Conflict of Interest

The authors have no conflicts of interest to declare that are relevant to the content of this chapter.

References

1. Tanaka N, Kanatani S, Tomer R, Sahlgren C, Kronqvist P, Kaczynska D, Louhivuori L, Kis L, Lindh C, Mitura P, Stepulak A, Corvigno S, Hartman J, Micke P, Mezheyeuski A, Strell C, Carlson JW, Fernandez Moro C, Dahlstrand H, Ostman A, Matsumoto K, Wiklund P, Oya M, Miyakawa A, Deisseroth K, Uhlen P (2017) Whole-tissue biopsy phenotyping of three-dimensional tumours reveals patterns of cancer heterogeneity. Nat Biomed Eng 1(10):796–806. https://doi.org/10.1038/s41551-017-0139-0
2. Weiss KR, Voigt FF, Shepherd DP, Huisken J (2021) Tutorial: practical considerations for tissue clearing and imaging. Nat Protoc 16(6):2732–2748. https://doi.org/10.1038/s41596-021-00502-8
3. Silvestri L, Costantini I, Sacconi L, Pavone FS (2016) Clearing of fixed tissue: a review from a microscopist's perspective. J Biomed Opt 21(8):081205
4. Costantini I, Cicchi R, Silvestri L, Vanzi F, Pavone FS (2019) In-vivo and ex-vivo optical clearing methods for biological tissues: review. Biomed Opt Express 10(10):5251–5267
5. Richardson DS, Lichtman JW (2015) Clarifying tissue clearing. Cell 162(2):246–257
6. Ueda HR, Erturk A, Chung K, Gradinaru V, Chedotal A, Tomancak P, Keller PJ (2020) Tissue clearing and its applications in neuroscience. Nat Rev Neurosci 21(2):61–79
7. Axer M, Amunts K (2022) Scale matters: the nested human connectome. Science 378(6619):500–504. https://doi.org/10.1126/science.abq2599
8. Di Meo D, Sorelli M, Ramazzotti J et al (2025) Quantitative cytoarchitectural phenotyping of deparaffinized human brain tissues. Commun Biol 8(1):1527. https://doi.org/10.1038/s42003-025-08887-y
9. Ando K, Laborde Q, Lazar A, Godefroy D, Youssef I, Amar M, Pooler A, Potier MC, Delatour B, Duyckaerts C (2014) Inside Alzheimer brain with CLARITY: senile plaques, neurofibrillary tangles and axons in 3-D. Acta Neuropathol 128(3):457–459. https://doi.org/10.1007/s00401-014-1322-y
10. Phillips J, Laude A, Lightowlers R, Morris CM, Turnbull DM, Lax NZ (2016) Development of passive CLARITY and immunofluorescent labelling of multiple proteins in human cerebellum: understanding mechanisms of neurodegeneration in mitochondrial disease. Sci Rep 6:26013
11. Costantini I, Mazzamuto G, Roffilli M, Laurino A, Maria Castelli F, Neri M, Lughi G, Simonetto A, Lazzeri E, Pesce L, Destrieux C, Silvestri L, Conti V, Guerrini R, Saverio Pavone F (2021) Large-scale, cell-resolution volumetric mapping allows layer-specific investigation of human brain cytoarchitecture. Biomed Opt Express 12(6):3684–3699
12. Pesce L, Laurino A, Scardigli M, Yang J, Boas DA, Hof PR, Destrieux C, Costantini I, Pavone FS (2022) Exploring the human cerebral cortex using confocal microscopy. Prog Biophys Mol Biol 168:3–9
13. Murray E, Cho JH, Goodwin D, Ku T, Swaney J, Kim SY, Choi H, Park YG, Park JY, Hubbert A, McCue M, Vassallo S, Bakh N, Frosch MP, Wedeen VJ, Seung HS, Chung K (2015) Simple, scalable proteomic imaging for high-dimensional profiling of intact systems. Cell 163(6):1500–1514
14. Liebmann T, Renier N, Bettayeb K, Greengard P, Tessier-Lavigne M, Flajolet M (2016) Three-dimensional study of Alzheimer's disease hallmarks using the iDISCO clearing method. Cell Rep 16(4):1138–1152
15. Hama H, Hioki H, Namiki K, Hoshida T, Kurokawa H, Ishidate F, Kaneko T, Akagi T, Saito T, Saido T, Miyawaki A (2015) ScaleS: an optical clearing palette for biological imaging. Nat Neurosci 18(10):1518–1529
16. Zhao S, Todorov MI, Cai R, Maskari RA, Steinke H, Kemter E, Mai H, Rong Z, Warmer M, Stanic K, Schoppe O, Paetzold JC, Gesierich B, Wong MN, Huber TB, Duering M, Bruns OT, Menze B, Lipfert J, Puelles VG, Wolf E, Bechmann I, Erturk A (2020) Cellular and molecular probing of intact human organs. Cell 180(4):796–812.e719. https://doi.org/10.1016/j.cell.2020.01.030
17. Schueth A, Hildebrand S, Samarska I, Sengupta S, Kiessling A, Herrler A, Zur Hausen A, Capalbo M, Roebroeck A (2023) Efficient 3D light-sheet imaging of very large-scale optically cleared human brain and prostate tissue samples. Commun Biol 6(1):170. https://doi.org/10.1038/s42003-023-04536-4
18. Ku T, Guan W, Evans NB, Sohn CH, Albanese A, Kim JG, Frosch MP, Chung K (2020) Elasticizing tissues for reversible shape transformation and accelerated molecular labeling. Nat Methods 17(6):609–613. https://doi.org/10.1038/s41592-020-0823-y

19. Shen FY, Harrington MM, Walker LA, Cheng HPJ, Boyden ES, Cai D (2020) Light microscopy based approach for mapping connectivity with molecular specificity. Nat Commun 11(1): 4632. https://doi.org/10.1038/s41467-020-18422-8
20. Scardigli M, Pesce L, Brady N, Mazzamuto G, Gavryusev V, Silvestri L, Hof PR, Destrieux C, Costantini I, Pavone FS (2021) Comparison of different tissue clearing methods for three-dimensional reconstruction of human brain cellular anatomy using advanced imaging techniques. Front Neuroanat 15:752234. https://doi.org/10.3389/fnana.2021.752234
21. Mitra S, Tomar PC (2021) Hybridoma technology; advancements, clinical significance, and future aspects. J Genet Eng Biotechnol 19(1):159. https://doi.org/10.1186/s43141-021-00264-6
22. Pesce L, Scardigli M, Gavryusev V, Laurino A, Mazzamuto G, Brady N, Sancataldo G, Silvestri L, Destrieux C, Hof PR, Costantini I, Pavone FS (2022) 3D molecular phenotyping of cleared human brain tissues with light-sheet fluorescence microscopy. Commun Biol 5(1): 447. https://doi.org/10.1038/s42003-022-03390-0
23. Costantini I, Ghobril JP, Di Giovanna AP, Allegra Mascaro AL, Silvestri L, Mullenbroich MC, Onofri L, Conti V, Vanzi F, Sacconi L, Guerrini R, Markram H, Iannello G, Pavone FS (2015) A versatile clearing agent for multimodal brain imaging. Sci Rep 5:9808
24. Choi SW, Guan W, Chung K (2021) Basic principles of hydrogel-based tissue transformation technologies and their applications. Cell 184(16):4115–4136. https://doi.org/10.1016/j.cell.2021.07.009
25. Costantini I (2023) A cellular resolution atlas of Broca's area. Sci Adv 9(41):eadg3844
26. Gorgolewski KJ, Auer T, Calhoun VD, Craddock RC, Das S, Duff EP, Flandin G, Ghosh SS, Glatard T, Halchenko YO, Handwerker DA, Hanke M, Keator D, Li X, Michael Z, Maumet C, Nichols BN, Nichols TE, Pellman J, Poline JB, Rokem A, Schaefer G, Sochat V, Triplett W, Turner JA, Varoquaux G, Poldrack RA (2016) The brain imaging data structure, a format for organizing and describing outputs of neuroimaging experiments. Sci Data 3:160044. https://doi.org/10.1038/sdata.2016.44
27. Gorgolewski KJ, Alfaro-Almagro F, Auer T, Bellec P, Capota M, Chakravarty MM, Churchill NW, Cohen AL, Craddock RC, Devenyi GA, Eklund A, Esteban O, Flandin G, Ghosh SS, Guntupalli JS, Jenkinson M, Keshavan A, Kiar G, Liem F, Raamana PR, Raffelt D, Steele CJ, Quirion PO, Smith RE, Strother SC, Varoquaux G, Wang Y, Yarkoni T, Poldrack RA (2017) BIDS apps: improving ease of use, accessibility, and reproducibility of neuroimaging data analysis methods. PLoS Comput Biol 13(3):e1005209. https://doi.org/10.1371/journal.pcbi.1005209
28. Sorelli M, Costantini I, Bocchi L, Axer M, Pavone FS, Mazzamuto G (2023) Fiber enhancement and 3D orientation analysis in label-free two-photon fluorescence microscopy. Sci Rep 13(1):4160. https://doi.org/10.1038/s41598-023-30953-w
29. Berg S, Kutra D, Kroeger T, Straehle CN, Kausler BX, Haubold C, Schiegg M, Ales J, Beier T, Rudy M, Eren K, Cervantes JI, Xu B, Beuttenmueller F, Wolny A, Zhang C, Koethe U, Hamprecht FA, Kreshuk A (2019) ilastik: interactive machine learning for (bio)-image analysis. Nat Methods 16(12): 1226–1232. https://doi.org/10.1038/s41592-019-0582-9
30. Amunts K, Lepage C, Borgeat L, Mohlberg H, Dickscheid T, Rousseau ME, Bludau S, Bazin PL, Lewis LB, Oros-Peusquens AM, Shah NJ, Lippert T, Zilles K, Evans AC (2013) BigBrain: an ultrahigh-resolution 3D human brain model. Science 340(6139):1472–1475. https://doi.org/10.1126/science.1235381
31. de Boer JF, Hitzenberger CK, Yasuno Y (2017) Polarization sensitive optical coherence tomography – a review [Invited]. Biomed Opt Express 8(3):1838–1873. https://doi.org/10.1364/BOE.8.001838
32. Slomianka L (2021) Basic quantitative morphological methods applied to the central nervous system. J Comp Neurol 529(4):694–756. https://doi.org/10.1002/cne.24976

Chapter 10

BRANT Pipeline to Map Neuronal Activations Across the Entire Brain: An Implementation Guide

Alessandra Franceschini

Abstract

Whole-brain pipelines are essential tools to study brain activity on a brain-wide scale. Currently, these pipelines, which combine different cutting-edge techniques—ranging from the use of transgenic animals to machine learning software—remain a niche area, limited to experts in the field. This chapter introduces BRANT (BRAin-wide Neuron quantification Toolkit), a scalable and user-friendly pipeline for three-dimensional (3D) analysis of whole-brain neuronal activation pattern that combines FosTRAP transgenic strategy, CLARITY tissue clearing protocol, high-resolution RAPID-enabled light-sheet fluorescent microscopy (LSFM), and 3D automated data analysis. By providing a detailed overview of its methodology and applications, this resource seeks to enhance accessibility and encourage the broader use of BRANT in whole-brain analysis.

Key words Whole-brain mapping, Immediate early genes, Clearing tissue protocols, Light sheet imaging, 3D analysis

1 Introduction

Understanding how neurons process and store information across the entire brain is a fundamental challenge in neuroscience. Over the years, many researchers faced this challenge limiting their studies on individual brain regions, thus failing to provide comprehensive explanations of brain mechanisms. Indeed, the study of single regions results in a fragmented and incomplete understanding of neural processes, making the interpretation of brain functions difficult. A classic example from traditional neuroscience is the study of memory, where researchers have predominantly focused on specific brain regions, such as the hippocampus for contextual component and the amygdala for emotional component. While these studies have provided valuable insights, they have also left unresolved many aspects of memory mechanisms and dynamics, highlighting the need for a more integrated, whole-brain approach. As a result, the need to fully comprehend brain mechanisms has driven researchers

Ludovico Silvestri (ed.), *Light Sheet Microscopy*, Neuromethods, vol. 229, https://doi.org/10.1007/978-1-0716-5100-1_10,

to extend their studies beyond isolated regions, shifting towards a whole-brain approach. Brain-wide neuronal activation mapping is a recently introduced technique that enables researchers to visualize and quantify neuronal activity across different regions, providing crucial insights into how neural circuits function, adapt, and respond to external stimuli.

The limited adoption of brain-wide pipelines is primarily due to significant technical challenges, including the complexity of large-scale data analysis, the large volume of data and the requirement for advanced microscopy techniques, demanding both the resolution and depth penetration necessary for whole-brain imaging. As a result, these methodological obstacles have largely restricted the use of such approaches to specialists in the field [1–7]. Although only a few pipelines are currently used, no single standardized method exists for whole-brain mapping. However, most techniques follow a common framework, utilizing IEG approaches to tag activated neurons in vivo and imaging them ex vivo using 3D optical tools such as LSFM. The integration of tissue clearing methods has significantly advanced brain-wide mapping by enabling the visualization of IEG^+ neurons across the entire brain. It is important to note that IEG-based transgenic tools are not standardized and should be selected based on specific research objectives. These tools can be categorized according to the localization of reporter expression, which may be either restricted to the nucleus or distributed throughout the entire neuron. When the fluorescence reporter labels the entire neuron, including axons and dendrites, large-scale brain-wide analysis becomes significantly more complex. This challenge arises because the fluorescent signal extends along neuronal processes, potentially interfering with detection and quantification algorithms, leading to the misidentification of activated neurons. Indeed, the widespread distribution of the signal in transgenic models requires subcellular resolution imaging to accurately distinguish neuronal nuclei from surrounding processes. This not only complicates image analysis but also dramatically increases data volume, making large-scale quantification computationally demanding. Unfortunately, current analytical tools are not yet optimized for efficiently handling terabyte-scale datasets, thereby limiting brain-wide pipelines to the use of IEG nuclear immunostaining.

BRANT is a novel pipeline for whole-brain mapping that leverages TRAP mice, CLARITY tissue clearing protocol, high-resolution LSFM, and terabyte-scale image processing [1] (Fig. 1). Compared to other brain-wide pipelines, BRANT uses a custom-made high-resolution LSFM integrated with a new autofocus system, called RAPID (Rapid Autofocus via Pupil-split Image phase Detection) that allows for a real-time correction of image defocus in any wide-field microscope [8, 9]. By applying high-resolution imaging across the entire brain, BRANT generates

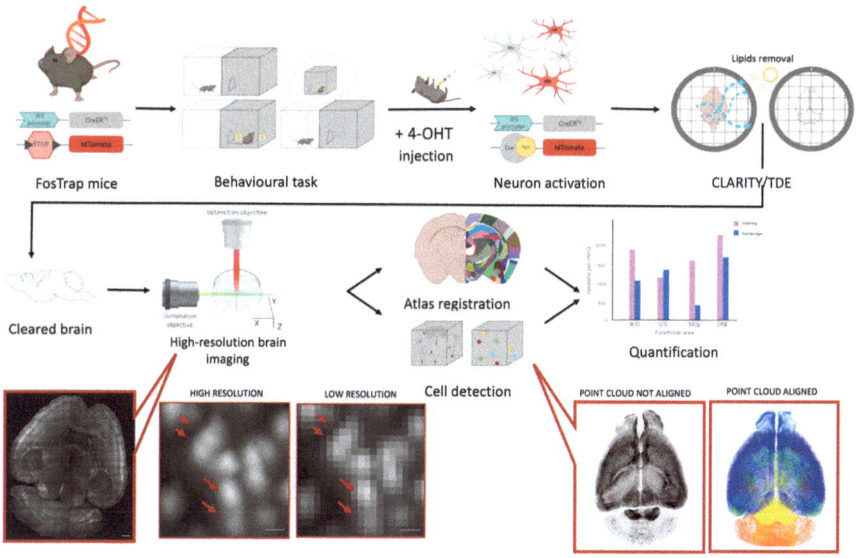

Fig. 1 BRANT pipeline. Schematic representation of the BRANT pipeline. FosTRAP mice are subjected to a behavioral task, followed by 4-OHT injection to induce permanent tagging of active neurons expressing tdTomato. After neuronal activation, brains are processed using CLARITY/thiodiethanol (TDE) for tissue clearing and lipid removal. Cleared brains are imaged with a custom high-resolution light-sheet fluorescence microscope equipped with orthogonal illumination and detection objectives. Representative images illustrate the benefit of high-resolution imaging for cell detection (red arrows) compared to low-resolution imaging. Acquired image stacks are registered to a reference brain atlas for spatial mapping, followed by automated cell detection and point cloud alignment. Quantitative analyses are then performed across brain regions to compare neuronal activation patterns between experimental conditions (e.g., training vs. home cage). (Adapted from Franceschini et al. [1])

terabytes of data per brain, necessitating a scalable analytical framework to efficiently process neuronal activity patterns in behaviorally relevant cohorts. Like many whole-brain pipelines, BRANT has been successfully used to investigate neuronal connectivity across different phases of fear memory, including memory encoding and recall, providing valuable insights into the neural circuits underlying learning and memory [1]. Thanks to its scalability and versatility, BRANT has the potential to be adapted for a wide range of applications, particularly in studies requiring the investigation of cellular dynamics across multiple brain regions. By providing high-resolution, large-scale insights into neuronal activity across the entire brain, BRANT could contribute to identifying disease-specific connectivity patterns, ultimately aiding in the development of more targeted therapeutic strategies.

2 A Practical Guide for the Use of BRANT Pipeline

Here, we provide a comprehensive and practical guide to using the BRANT pipeline, detailing each step to facilitate its implementation. This guide will cover the sample preparation, imaging process, data acquisition, and analysis workflow, ensuring that readers can effectively apply BRANT for whole-brain mapping.

2.1 Sample Preparation and Clearing Protocol

One of the principal goals of studying whole-brain mechanisms is to understand which brain circuits are responsible for a specific behavioral readout. Indeed, BRANT was applied to study the neuronal circuits underlying fear memory in both sexes. Many of these studies are first conducted in animal models, commonly in rodent models (e.g., rats and mice), to provide essential insights into brain function. New discoveries from these preclinical studies will then serve as the foundation for clinical research or further investigations in humans, helping to bridge the gap between basic neuroscience and medical applications. One of the key points for the whole-brain studies is the brain structure preservation. The need to preserve the entire 3D anatomical structure led the researchers to the development of clearing tissue protocols. Transparency offers direct optical access to large specimens (e.g., the brain), eliminating the need for physical sectioning. More broadly, preserving whole brains instead of analyzing smaller fragments is essential to gain a more comprehensive understanding of neuronal mechanisms. The principle behind clearing protocols is to make the brain transparent by equalizing the refractive index (RI) of its solid and liquid components, thereby matching the RI of the tissue with that of the surrounding medium [10, 11]. To enable whole-brain reconstruction, the CLARITY tissue clearing protocol was integrated into the experimental pipeline [12]. This method was specifically selected over other clearing techniques due to its ability to effectively preserve the endogenous TdTomato fluorescent reporter in TRAP mice. Moreover, the use of a solution containing 40% 2,2′-thiodiethanol (TDE) as RI-matching solution induces a slight expansion of the brain tissue, which may represent an advantage when investigating small components, such as neuronal processes, making them easier to resolve [13, 14]. Furthermore, the combination of the 40% TDE solution with high-resolution imaging facilitates the performance of analysis algorithms, improving their ability to discriminate between neuronal cell bodies, processes, and blood vessels.

When applying this pipeline, the process begins with a critical step: brain extraction following mouse perfusion. It is essential to carefully remove the brain from the cranial cavity without causing any damage, as cuts or incisions could compromise the subsequent three-dimensional reconstruction. Immediately after extraction, the brains are immersed in a 4% paraformaldehyde (PFA) solution to preserve cellular and molecular structures, thereby preventing tissue degradation and maintaining integrity. PFA acts as a fixative by cross-linking proteins and stabilizing the brain's architecture for further analysis. Nevertheless, it is important to note that prolonged exposure to PFA can increase tissue autofluorescence, which may interfere with downstream imaging. To avoid this, fixed brains should be thoroughly washed with phosphate-buffered saline (PBS) after overnight fixation. Following this step, the samples are incubated in a hydrogel solution containing a thermal initiator. In this phase, the exclusion of oxygen and the application of nitrogen and elevated temperatures (37 °C) are required to trigger hydrogel polymerization, which results in the formation of a mesh able to cross-link proteins and nucleic acids. This step is crucial and highly delicate, since errors in the polymerization process, which are often caused by the presence of oxygen, may prevent the proper formation of the hydrogel mesh. Therefore, the tissue structure may collapse during subsequent treatment with strong detergents. Indeed, a fundamental step in clearing protocols is lipid removal. Lipids are not cross-linked to the hydrogel mesh, and the treatment with strong basic detergents—such as a solution containing sodium dodecyl sulfate (SDS) and boric acid—efficiently removes lipids from the tissue. This process makes the hydrogel–tissue hybrid transparent. The final step before imaging is the immersion of the sample in the TDE 40% solution to homogenize the RI of the dry component with that of the surrounding medium, thereby achieving high sample transparency for high-resolution imaging.

2.2 High-Resolution Light-Sheet Microscopy

A detailed description of LSFM is beyond the scope of this chapter. Readers interested in the technical principles of LSFM are referred to other chapters of this book. The aim of this paragraph is to briefly highlight how LSFM has become the technique of choice for 3D reconstructions and how it has been selected and integrated within the BRANT pipeline. The reasons LSFM is preferred for large-scale studies are fast volumetric imaging, low photobleaching, high flexibility, and compatibility with clearing protocols. One of the most notable strengths of LSFM is the high imaging speed, achieved through the simultaneous illumination of multiple points within a single optical plane. This feature speeds up the acquisition of 3D datasets compared to confocal or two-photon microscopy, enabling the reconstruction of an entire mouse brain within a time frame ranging from several hours to just minutes, depending on the required resolution. Moreover, since LSFM excites only the focal plane (and not the entire sample volume), it minimizes photodamage and bleaching. Equally important is the high versatility of LSFM, which allows users to balance imaging speed and resolution according to experimental needs. For instance, low-resolution

LSFM is widely used in combination with cFos nuclear staining for brain-wide cell counting, resulting in datasets of manageable size [15–17]. Finally, the use of clearing protocols allows light-sheets to deeply penetrate with minimal scattering, avoiding the physical sectioning of the sample.

Taken together, these properties make LSFM one of the preferred imaging modalities for 3D reconstruction of large biological samples. Its ability to image entire, optically cleared brain samples rapidly and with minimal photobleaching and photodamage has positioned LSFM as a key tool for large-scale anatomical studies [9, 18–21].

The integration of a custom high-resolution LSFM in the BRANT pipeline enabled the imaging and analysis of large cohorts of samples. Indeed, the use of the TRAP mouse line needed a resolution on the order of microns to ensure optimal image quality across large, cleared brains, thereby maximizing the disentanglement of dense environments for the following automated analysis. However, the inability to maintain high resolution, especially when imaging goes deep inside the sample, led to the integration of a real-time image-based autofocus system, known as RAPID [8, 9]. To realize sharp images to be included in the BRANT pipeline, there are a few tips worth mentioning that can be potentially useful for anyone who wants to approach to it. First, the use of an immersion detection objective specifically corrected for clearing solutions (XLPLN10XSVMP 10X NA 0.6, Olympus) helps minimize optical aberrations. The use of an index matching solution with an RI different from that of water or air, such as the use of TDE, requires a specific objective incorporating a ring for RI correction. These objectives are designed to compensate for optical distortions caused by the different refractive index of the specific clearing agent, ensuring uniform resolution and image quality throughout the sample. Another key consideration relates to the working distance of the objective. The lens has a working distance limited to 8 mm, which makes it unsuitable for imaging an intact mouse brain in a single acquisition. As a result, brains were imaged in two separate halves, with an overlap region of approximately 1 mm to facilitate subsequent alignment and stitching. Third, the custom high-resolution LSFM system delivers exceptional optical performance, achieving a resolution of 1.35 μm in the imaging plane and 8.6 μm along the *z*-axis [14]. Voxel dimensions are finely tuned to $0.65 \times 0.65 \times 2$ μm along the *x*, *y*, and *z* axes, respectively, ensuring high spatial precision. However, due to the limited field of view—approximately 1.3×1.3 mm^2—whole-brain datasets are acquired as a mosaic of partially overlapping image stacks. This tiling strategy guarantees full coverage of the sample while preserving the high resolution required for detailed structural analysis. Taken together, these characteristics define a robust and versatile imaging pipeline, ideally suited for large-volume, high-resolution brain mapping. The combination of optical precision and systematic coverage makes it possible to reconstruct brain-wide structures with a level of detail that supports both qualitative and quantitative analysis.

2.3 Image Processing, Cell Quantification, and Data Analysis

Often, the most challenging part of brain-wide engram analysis is the conversion of raw images, identified as a matrix of gray intensity values, into quantitative datasets providing relevant biological information. This process involves three different phases: (1) automated cell localization in large-scale datasets, (2) spatial registration to a reference atlas, and (3) interpretation of data on a brain-wide scale (Fig. 2). One of the main cores of the BRANT pipeline, together with high-resolution imaging, is a robust and scalable framework for image processing and for data analysis of whole-brain neuronal activations at single-cell resolution. After imaging, the raw dataset per brain often exceeds 16 terabytes and therefore needs to be compressed by a factor of 20 using a 16-bit lossy JPEG-2000 format [1]. These compressed image volumes are then stitched using ZetaStitcher, a custom-built software tool that allows efficient and seamless access to high-resolution data in specific subvolumes through its VirtualFusedVolume application programming interface (API) [1]. Working directly with uncompressed raw data is not recommended in this pipeline, as it substantially increases disk space requirements and leads to significantly slower data access and processing times, particularly when handling large, tiled image volumes. Although high-resolution imaging of the entire brain poses significant challenges for data handling and visualization, it is essential for accurately distinguishing densely packed structures such as axons, dendrites, and blood vessels. This level of detail necessitates the use of machine learning-based approaches for effective image analysis and feature extraction. BRANT uses bcFind, a deep-learning-based tool built on a U-Net architecture trained to isolate neuronal somata from surrounding structures such as dendrites, axons, and vasculature. Unlike traditional threshold-based methods, bcFind transforms raw images into a simplified representation where only fluorescent cell bodies are preserved, enabling precise quantification even in densely labeled environments [22]. On the image processing side, ZetaStitcher makes it possible to access small parts of very large brain images, allowing efficient and targeted analysis. In addition, the deep learning tool bcFind improves the accuracy of cell detection and performs better than traditional methods like ClearMap, which was developed for cFos nuclear staining. Indeed, ClearMap is designed to identify spherical objects and performs best with nuclear staining, where cell bodies have a more defined, compact shape. Another key component of the BRANT pipeline is the registration process using a single channel. After cell detection, the spatial coordinates of each neuron are aligned to the Allen

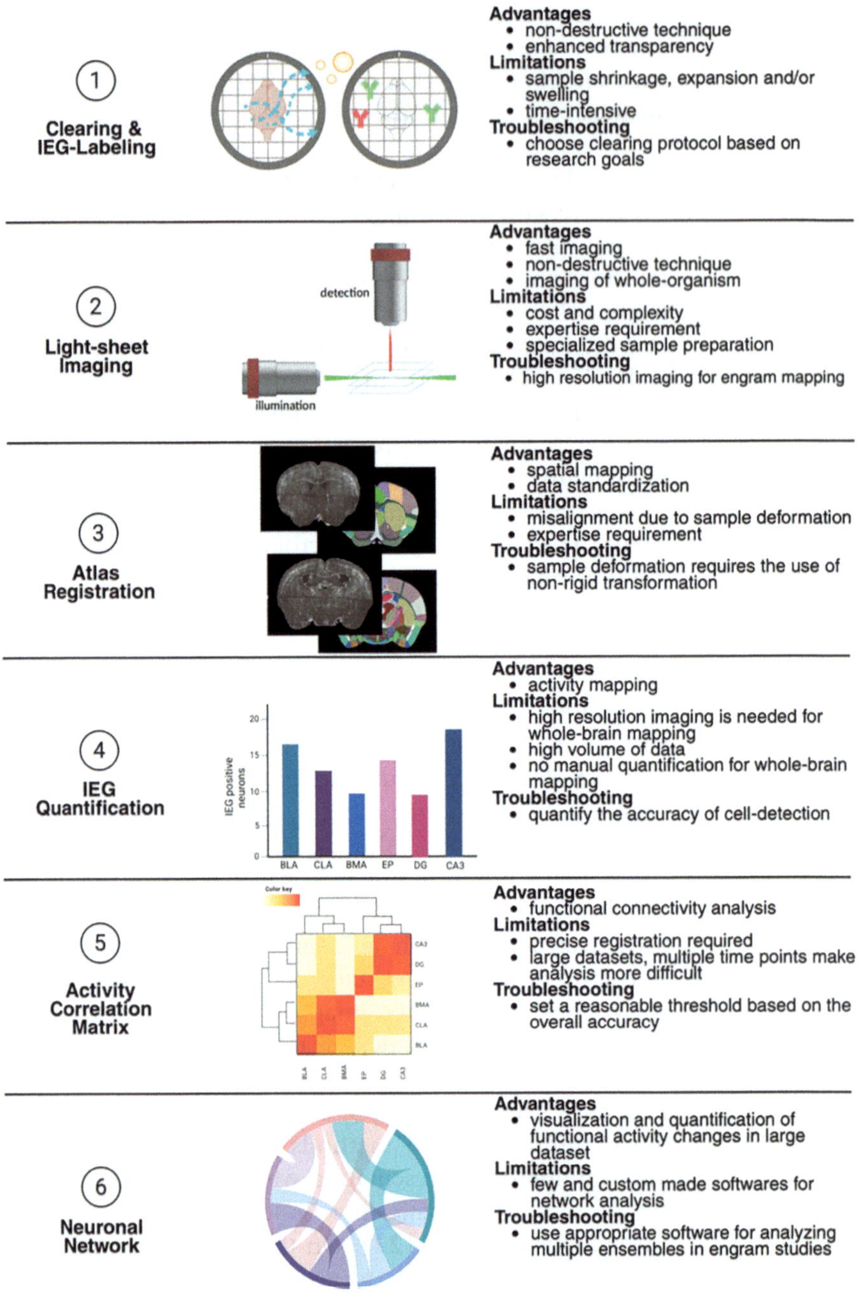

Fig. 2 Scheme of 3D-automated analysis pipeline used for engram mapping. This pipeline is composed of many steps; entire brains are made transparent using clearing tissue protocols and a selected IEG is labelled across the entire neuron. By using specific imaging techniques, an entire brain is reconstructed. All imaged brains are registered to a referred brain atlas to quantify the number of positive IEG neurons in each brain region. IEG quantifications are cross-correlated among brain regions to create a correlated activity matrix and consequently, a functional connectivity pattern. (Adapted from Franceschini et al. [23])

Brain Atlas using a single fluorescence channel, adjusted through gamma transformation. This approach eliminates the need for dual-channel imaging while still achieving a spatial accuracy of approximately 300 μm, which is sufficient for assigning cells to large-scale anatomical regions. Detected neurons are then grouped into 48 macro-regions, enabling robust comparisons between experimental groups [1]. Regional cell counts can be used to analyze the activity patterns that differ the most between experimental groups, using tools like partial least squares analysis (PLS) [7]. Finally, Pearson correlation in activation between different regions, calculated using each animal as an experimental point, can be used to infer functional connectivity, enabling the further use of network neuroscience analysis tools [1]. The streamlined computational workflow presented here allows for high-throughput analysis across different timepoints and conditions, supporting statistically sound evaluations of brain-wide activity patterns. By combining resolution, scalability, and automation, BRANT establishes a new benchmark for large-scale, three-dimensional mapping of neuronal activation.

2.4 Application

The BRANT pipeline has been effectively applied to investigate the time evolution of fear memory across the entire brain. Designed for whole-brain mapping, it is optimized for the comparison of multiple behaviorally relevant cohorts and supports the detailed study of memory-related processes on a brain-wide scale. Beyond its specific utility in memory, BRANT's combination of high-resolution imaging and automated 3D analysis makes it especially powerful for large-scale studies of behavioral neuroscience, neurological disorders, and a wide range of applications in systems neuroscience since all these use cases are not confined to a particular area but rather distributed across the entire brain. For the investigation of neurological disorders, the BRANT capacity to reveal alterations in brain-wide activity patterns provides valuable insights into the disrupted neural dynamics underlying various pathological conditions. Moreover, the versatility of this pipeline extends its applicability to a broader range of research questions, beyond traditional neuroscience, such as developmental biology, systems biology, oncology, and histology. As a result, BRANT offers a flexible and scalable platform that can be adapted to diverse experimental contexts, making it a valuable resource across multiple domains of medical and biomedical research.

2.5 Strengths and Limitations of the Pipeline

The BRANT pipeline offers a robust and scalable solution to map whole-brain neuronal activity at single-cell resolution. Its core strengths lie in the combination of fast and high-resolution imaging along with automated 3D analysis. As previously outlined in the preceding paragraphs, maintaining high image quality across large tissue volumes is made possible by the integration of RAPID.

Furthermore, BRANT is engineered to efficiently manage massive datasets. Thanks to efficient data compression and smart subvolume access, it can handle large TB-scale datasets per brain, enabling navigation and analysis of complex brain-wide information. BRANT also incorporates a deep-learning-based cell detection algorithm that provides high accuracy and robustness, particularly with large-scale studies. This is particularly advantageous when working with complex labeling strategies, such as the TRAP or ArcCreER^{T2} mouse model. Importantly, BRANT is based on fully open-source software available on GitHub and can thus be used, adapted, and improved by the entire community in a collaborative spirit. Despite all the many advantages offered by this pipeline, the use of high-resolution across the entire sample requires specialized equipment, large storage capacity, and processing times, which may limit accessibility for smaller laboratories. Additionally, deep-learning software involves high-performance computing resources and dedicated technical expertise. Another challenge lies in registration accuracy due to tissue deformation introduced using clearing protocols. Although the pipeline achieves a spatial registration accuracy of approximately 300 μm, this resolution may not suffice for analyses requiring finer anatomical discrimination. Furthermore, while BRANT is optimized for TRAP-labeled neurons with whole-cell fluorescence, its performance may vary with other labeling strategies, particularly those with lower signal-to-noise ratios or limited spatial distribution.

In conclusion, BRANT stands out as a cutting-edge platform that enables high-resolution, large-scale, and reproducible brain-wide analysis. Its combination of technical innovation, open accessibility, and adaptability to diverse experimental contexts makes it a powerful resource for advancing systems neuroscience.

Conflict of Interest

The authors have no conflicts of interest to declare that are relevant to the content of this chapter.

References

1. Franceschini A et al (2023) Brain-wide neuron quantification toolkit reveals strong sexual dimorphism in the evolution of fear memory. Cell Rep 42:112908
2. Guenthner CJ, Miyamichi K, Yang HH, Heller HC, Luo L (2013) Permanent genetic access to transiently active neurons via TRAP: targeted recombination in active populations. Neuron 78:773–784
3. Renier N et al (2014) iDISCO: a simple, rapid method to immunolabel large tissue samples for volume imaging. Cell 159:896–910
4. Roy DS et al (2022) Brain-wide mapping reveals that engrams for a single memory are distributed across multiple brain regions. Nat Commun 13:1799

5. Vetere G et al (2017) Chemogenetic interrogation of a brain-wide fear memory network in mice. Neuron 94:363–374.e4
6. Vousden DA et al (2015) Whole-brain mapping of behaviourally induced neural activation in mice. Brain Struct Funct 220:2043–2057
7. Wheeler AL et al (2013) Identification of a functional connectome for long-term fear memory in mice. PLoS Comput Biol 9: e1002853
8. Müllenbroich MC et al (2015) Comprehensive optical and data management infrastructure for high-throughput light-sheet microscopy of whole mouse brains. Neurophoton 2:041404
9. Silvestri L et al (2021) Universal autofocus for quantitative volumetric microscopy of whole mouse brains. Nat Methods 18:953–958
10. Richardson DS, Lichtman JW (2015) Clarifying tissue clearing. Cell 162:246–257
11. Tuchin VV (1997) Light propagation in tissues with controlled optical properties. J Biomed Opt 2:401
12. Chung K et al (2013) Structural and molecular interrogation of intact biological systems. Nature 497:332–337
13. Costantini I et al (2015) A versatile clearing agent for multi-modal brain imaging. Sci Rep 5:9808
14. Di Giovanna AP et al (2019) Tailored sample mounting for light-sheet fluorescence microscopy of clarified specimens by polydimethylsiloxane casting. Front Neuroanat 13:35
15. Renier N et al (2016) Mapping of brain activity by automated volume analysis of immediate early genes. Cell 165:1789–1802
16. Susaki EA et al (2014) Whole-brain imaging with single-cell resolution using chemical cocktails and computational analysis. Cell 157:726–739
17. Ye L et al (2016) Wiring and molecular features of prefrontal ensembles representing distinct experiences. Cell 165:1776–1788
18. Chen Y et al (2024) Low-cost and scalable projected light-sheet microscopy for the high-resolution imaging of cleared tissue and living samples. Nat Biomed Eng. https://doi.org/10.1038/s41551-024-01249-9
19. Dodt H-U et al (2007) Ultramicroscopy: three-dimensional visualization of neuronal networks in the whole mouse brain. Nat Methods 4:331–336
20. Keller PJ, Dodt H-U (2012) Light sheet microscopy of living or cleared specimens. Curr Opin Neurobiol 22:138–143
21. Ueda HR et al (2020) Whole-brain profiling of cells and circuits in mammals by tissue clearing and light-sheet microscopy. Neuron 106:369–387
22. Frasconi P et al (2014) Large-scale automated identification of mouse brain cells in confocal light sheet microscopy images. Bioinformatics 30:i587–i593
23. Franceschini A, Jin M, Chen CW, Silvestri L, Mastrodonato A, Denny CA (2025) Brain-wide immunolabeling and tissue clearing applications for engram research. Neurobiol Learn Mem. Mar;218:108032. https://doi.org/10.1016/j.nlm.2025.108032. Epub 2025 Feb 6. PMID: 39922482

Chapter 11

High Spatiotemporal Resolution Functional Imaging of the Entire Larval Zebrafish Brain with Two-Photon Light-Sheet Fluorescence Microscopy

Lapo Turrini, Giuseppe de Vito, and Francesco Saverio Pavone

Abstract

Light-sheet fluorescence microscopy (LSFM) has revolutionized brain imaging by enabling high-speed volumetric recording of large translucent samples, with zebrafish larvae being a prime model for whole-brain functional studies. Among its variations, two-photon (2P) LSFM stands out for its use of pulsed infrared lasers, offering deeper imaging with reduced scattering and absorption, minimal autofluorescence, and avoidance of unwanted visual responses in live animals. First introduced for functional imaging in 2011, 2P LSFM has since evolved with manifold advancements broadening its applications. Despite these advantages, challenges such as lower signal-to-noise ratios, photodamage thresholds, and the need for complex optical setups remain. In this chapter, we discuss the improvements brought by 2P LSFM and explore key innovations in this technique that address the aforementioned limitations, including optimized laser pulse frequencies, polarization techniques, and advanced beam-splitting methods to enhance fluorescence efficiency and imaging quality. We also examine the challenges associated with the generation of large datasets and introduce data processing pipelines commonly employed for efficient handling and analysis. Finally, we present a case study on the application of 2P LSFM to investigate the propagation of epileptic activity in zebrafish larvae. This approach enables real-time visualization of spatiotemporal patterns of seizure propagation across the brain, offering critical insights into the mechanisms of neuronal synchronization and circuit dysfunction. By bridging cellular and whole-brain scales, this example highlights the transformative potential of 2P LSFM in advancing our understanding of large-scale neuronal activity and complex brain disorders.

Key words Zebrafish larva, Functional imaging, Light-sheet fluorescence microscopy, Two-photon excitation, Calcium imaging

1 Introduction

Light-sheet fluorescence microscopy (LSFM) [1] has become a pivotal tool in brain imaging due to its unique design, which combines optical sectioning with parallel acquisition within each

Lapo Turrini and Giuseppe de Vito contributed equally to the writing of this chapter.

Ludovico Silvestri (ed.), *Light Sheet Microscopy*, Neuromethods, vol. 229, https://doi.org/10.1007/978-1-0716-5100-1_11,

Table 1
Comparison between one and two-photon light-sheet fluorescence microscopy

Imaging technique	Light source	Illumination mode	Cost	Most frequent application	Typical limiting factor in volumetric speed
1P LSFM	Continuous-wave laser, usually in the visible range	Static or dynamically scanned	€	Structural/functional imaging	Camera frame rate and/or refocusing inertia
2P LSFM	Pulsed infrared laser	Dynamically scanned	€€€	Functional brain imaging and structural imaging in opaque samples	Photodamage/ SNR

frame, enabling high-speed volumetric imaging of large translucent samples [2–4]. LSFM was the first technique allowing the functional imaging of an entire vertebrate brain, using the transparent zebrafish (*Danio rerio*) larvae as a model [5]. This technique provided new insights into the integrated dynamics of neural circuits [6–10] through the simultaneous imaging of large neuronal populations using genetically encoded fluorescent calcium indicators.

Two-photon (2P) LSFM is a specialized variant of LSFM that uses pulsed infrared lasers for 2P excitation [11, 12], contrasting with the continuous-wave lasers used in traditional one-photon (1P) LSFM (Table 1). The first pioneering work presenting 2P LSFM implementations used static light sheets generated by cylindrical lenses [13], achieving anatomical (i.e., not functional) imaging in live organisms (*C. elegans*). However, in 2011, a more efficient 2P LSFM setup using a digitally scanned light sheet (a modality where the beam is rapidly scanned across the plane to produce the light sheet) was introduced by Truong and colleagues [14], enabling in vivo structural imaging in *D. melanogaster* and functional imaging of the beating heart in *D. rerio.* A second functional application of 2P LSFM was presented by Wolf and colleagues [15] in 2015, who successfully recorded calcium transients in the whole brain of the zebrafish larva. This work marked the beginning of 2P LSFM as a suitable tool for brain functional imaging [16–22]. Subsequent advancements in 2P LSFM included variations like Bessel [23–25] and Airy beams [26], multiview detection [27], multicolor excitation [28] and its combination with structured illumination [24], depth of focus extension [29, 30], super-resolution [31, 32], and light-field microscopy [18], which allowed to expand the realm of application of 2P LSFM.

The shift from 1P to 2P illumination provides several advantages. Infrared (IR) light, used in 2P LSFM, has reduced scattering and absorption [33] in biological tissues, allowing for deeper imaging while also minimizing autofluorescence [14, 34, 35].

Unlike 1P LSFM, which uses visible light that can trigger unwanted visual responses in animals through the activation of retino-recipient neurons (as noted by Wolf and collaborators [15]), the IR light of 2P LSFM avoids such effects, being invisible to most vertebrates [36], including zebrafish [37].

The use of infrared light in 2P LSFM thus eliminates the issue of undesired visual stimulation, making it a key advantage over 1P LSFM for functional whole-brain imaging performed in living animals. This is the reason why in vivo 2P LSFM is currently applied almost exclusively to zebrafish imaging (or similar animals, such as medaka fish [25], with the rare exception of *D. melanogaster* [38]). This capability makes 2P LSFM especially suitable for functional studies where visual stimulus independence is critical, such as experiments involving target-driven fictive navigation [17], sleep/wake cycles [39, 40], or pathological brain states characterized by hyperexcitability [21]. Additionally, 2P excitation enables more reliable behavioral data collection alongside neuronal imaging in partially restrained animals. While IR is not detected by the larval visual system, solutions have been developed to limit the exposure of the retina to prolonged direct IR laser illumination, which could cause tissue damage. One approach involves using opaque masks to shield the eyes [19, 41, 42]. However, this method can prevent imaging of the brain portion located between the eyes. An alternative solution employs an additional laser beam entering from the front to scan the interocular brain region while avoiding this area with the lateral illumination beam(s) [43]. Although effective, this approach significantly increases the complexity of the experimental setup.

Furthermore, 2P excitation facilitates better separation of excitation and emission wavelengths, due to the blue-shifted nature of the elicited fluorescent light. Additionally, 2P excitation spectra are typically broader compared to their 1P counterparts, offering greater flexibility in selecting an appropriate illumination wavelength. This flexibility is particularly valuable in integrated systems, such as when combining LSFM with optogenetic stimulation in living animals [22], where it is essential to use a non-overlapping wavelength for optogenetic actuator stimulation.

Finally, the reduced absorption and scattering of infrared light, combined with the quadratic dependence of the absorption on the illumination intensity, helps mitigate shadowing artifacts—a common challenge in LSFM [44]. It also reduces background fluorescence [45] and sample-induced aberrations [34], thereby enhancing image uniformity and contrast.

These advantages come with some challenges. 2P excitation has a lower cross-section compared to 1P absorption, which can reduce the signal-to-noise ratio (SNR) at fixed illumination power levels [46]. Therefore, since the illumination power maximization is often constrained by the photodamage threshold, maximization of fluorescence excitation/collection efficiency in 2P LSFM is crucial. Moreover, 2P LSFM setups require bulkier and more expensive pulsed infrared lasers (often employing Glan–Thompson polarizers to control beam power) compared to continuous-wave visible lasers used in 1P LSFM. Light-sheet generation in 2P LSFM typically relies on dynamic methods (using galvanometric scanner mirrors) rather than static light sheets because of the need for high spatial and temporal concentration of the illumination power. Indeed, to concentrate the illumination power in space, it is quite common to observe 2P LSFM setups that use two illumination objectives to cover a field size which is more commonly covered by only one illumination objective in 1P LSFM. In addition, pulse compressors are often used to compensate for the group delay dispersion (GDD) of optical pulses caused by the microscope optics, thus ensuring effective 2P excitation. All these aspects increase the setup complexity.

Optimizing laser pulse frequency can significantly enhance signal generation while reducing photodamage [47, 48]; hence, pulsed lasers with low repetition rate and high peak power are preferable over the most commonly used 80-MHz repetition rate pulsed lasers. Additionally, our group demonstrated that exploiting polarization-dependent spatial anisotropy in light absorption and emission can increase the SNR in 2P LSFM [49]. Specifically, the absorption probability of excitation photons is higher if the direction of the electric field of these photons (i.e., the light beam polarization) is parallel to the absorption transition dipole of the fluorophore. This results in photoselection of the fluorophore population, meaning that, when using linearly polarized light to illuminate randomly oriented fluorophores, those with a specific spatial orientation will be more represented in the subpopulation of the excited molecules. After excitation, rotational diffusion due to Brownian motion will partially randomize the angular distribution of the excited fluorophores. However, for larger fluorophores in a biological environment, complete randomization may not occur within the excited fluorophore lifetime. Fluorescence emission from an excited fluorophore occurs preferentially along a plane perpendicular to the emission transition moment. Since the relative orientation of the absorption and emission dipole moments is fixed for a given molecule (often parallel), we can control the preferential direction of the fluorescence light emission by adjusting the polarization orientation of the linearly polarized excitation light. This effect is also observed in 1P confocal and 2P microscopy [50, 51] but is particularly important in LSFM due to the orthogonal arrangement of the illumination and detection objectives [52]. This arrangement allows tuning of the illumination light polarization to maximize fluorescence generation toward the detection objective, thereby increasing photon collection efficiency [53]. These considerations are even more critical for 2P LSFM compared to 1P LSFM, as the acquisition speed of 2P LSFM is often constrained by the SNR. Furthermore, due to the quadratic nature of the 2P excitation process, the absorption probability in this case depends on the fourth power of the cosine of the angle between the polarization of the excitation electric field and the fluorophore absorption dipole moment (compared to the second power in 1P excitation) [49]. This further amplifies the photoselection effect in 2P LSFM.

In a second work from our group [21], we described other expedients which can be used to increase the SNR in 2P LSFM. One of them is the use of polarizing beam splitters in place of non-polarizing beam splitters, combined with an electro-optic modulator (EOM) for fast switching of the polarization plane orientation of the illumination light. Beam splitters are frequently used in LSFM to feed light in multiple (usually two) illumination objectives. With the continuous-wave lasers employed in 1P LSFM, the use of simple non-polarizing beam splitters is effective. However, in 2P microscopy, a non-polarizing beam splitter halves the peak pulse power, while preserving the illumination time. On the contrary, the use of a polarizing beam splitter preserves the pulse peak power while halving the illumination time. In this way, due to the quadratic dependence of the 2P absorption probability on the peak power density (in contrast with the linear dependence of the signal on the integration time), the generated fluorescence signal with a polarizing beam splitter is double that obtained with a non-polarizing beam splitter.

In contrast to the illumination path, the detection arm of a 2P LSFM setup does not differ significantly from its 1P counterpart. However, using high-numerical aperture liquid-immersion objectives can significantly enhance signal collection and improve overall image quality. However, the use of liquid-immersion objectives moved at high frequency (usually >2 Hz) to perform functional volumetric imaging can introduce pressure waves which, impinging on the sample, completely disrupt the imaging process. In this case, a method of remote focusing (e.g., through the use of an electro-tunable lens [21, 54]), which allows the detection objective to remain stationary, is preferable.

2 An Effective Two-Photon Light-Sheet Fluorescence Setup

To achieve fast, high-resolution, noninvasive functional imaging of the entire larval zebrafish brain, we have developed a 2P digitally scanned LSF microscope [21] (Fig. 1). The system (Fig. 1a, c) uses

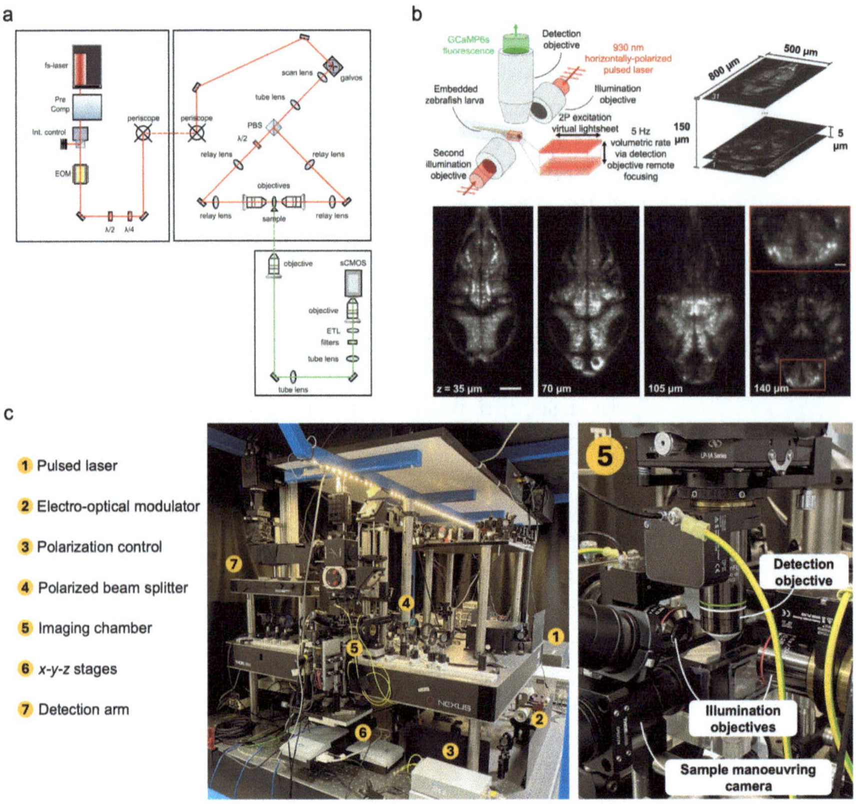

Fig. 1 An optimized 2P LSF setup for fast, whole-brain imaging of the larval zebrafish brain at high spatial resolution. (**a**) Schematic of the custom-built two-photon light sheet fluorescence microscopy system. The scheme includes key components: fs-laser (femtosecond laser), Pre-Comp (GDD precompensation unit), Int. control (power intensity control unit), EOM, λ/2 (half-wave plate), λ/4 (quarter-wave plate), galvos (hybrid galvanometric mirror assembly), PBS (polarizing beam splitter), and ETL. The excitation beam path is depicted by red lines, and the detection beam path by green lines. Vertical beam paths are indicated with dashed segments. (**b**) (*Top*) Close-up schematic illustrating the axial scanning direction, the polarization plane of the excitation light (double-headed arrows), and the geometric orientation of the larva and objectives. The schematic also depicts the key parameters of the volumetric acquisition process (right). (*Bottom*) Four transverse sections displaying temporal average intensity projections at varying dorso-ventral depths of the larva (depths indicated on each panel relative to the dorsal surface). Scale bar: 100 μm. The red box highlights a magnified region of the telencephalon, allowing visualization of single-neuron resolution. Scale bar for magnified view: 20 μm. (**c**) Images of the custom 2P LSFM setup with key components highlighted. (Panels **a**, **b** were adapted with permission from [21], © The Optical Society)

a tunable titanium–sapphire pulsed laser (Chameleon Ultra II, Coherent) as the excitation source, typically operating at a wavelength of 930 nm. To counteract the GDD caused by the optical components and the long path of the focused light in the water-filled imaging chamber, we incorporated a pulse compressor unit (PreComp, Coherent) immediately after the laser aperture. As outlined in the Introduction, compensating for GDD ensures the photons of different frequencies belonging to the same laser pulse reach the sample with minimal temporal dispersion, thus enhancing the probability of a two-photon excitation and improving the SNR. Following compression, the laser beam power is modulated using a half-wave plate and a Glan–Thompson polarizer, which diverts a controllable portion of the beam into a beam dump. The beam then passes through an electro-optical modulator (EOM, 84502050006, Qioptiq), which alternates the polarization state between two orthogonal modes at 100 kHz. A quarter-wave plate after the EOM pre-compensates polarization distortions, followed by a half-wave plate that aligns the beam polarization either parallel or perpendicular to the optical table (the significance of these steps will be discussed later). The beam is then directed by a series of steering mirrors through a periscope and to a hybrid system of galvanometric mirrors. One of these is a fast resonant mirror (CRS-8 kHz, Cambridge Technology) scanning the beam along the fronto-caudal axis of the larva at 8 kHz, thus creating a virtual light sheet. The other galvanometric mirror (6215H, Cambridge Technology) operates in closed-loop mode, scanning the light sheet along the dorso-ventral axis of the larva at frequencies up to 5 Hz. After the galvanometric mirrors, the laser beam is routed through a telescope consisting of a scan lens (50 mm focal length) and a tube lens (75 mm). Next, a polarizing beam splitter directs the laser light to either of the two excitation arms, based on the instantaneous polarization state determined by the EOM. As anticipated, this step is critical, as halving the peak power in each excitation path would otherwise halve the emitted fluorescence signal (due to the quadratic dependency of two-photon excitation on illumination intensity versus the linear dependency on illumination time). A half-wave plate then rotates the light polarization in one of the two arms, ensuring it to be aligned parallel to the optical table surface for both beams, which is essential for optimal performance [49], as outlined in the Introduction. Each excitation arm contains a pair of relay lenses (250 mm and 200 mm focal length, respectively) that, together with the scan lens/tube lens pair, magnify the beam diameter by 1.2 times before directing it to the two dry excitation objectives (XLFLUOR4X/340/0.28, Olympus), positioned laterally to the larva. This magnification results in underfilling of the objective pupils, producing an elongation of the axial point spread function (PSF) of the focused beam that is instrumental to the light-sheet formation. Owing to the direct relationship existing between the Rayleigh length and the thickness of a Gaussian-distributed beam, a less focused beam will have a length covering the entire width of the larval brain (approximately 400–500 μm at 6 days post fertilization), but with increased light-sheet thickness that would compromise optical sectioning. The solution we employed consists in using two light sheets, each covering half the brain width, with an axial full width at half maximum comparable to neuron size. The focused excitation light from both objectives is directed into a custom water-filled sample chamber with thin glass walls (0.17 mm) and an open top for access by a water-immersion detection objective (XLUMPLFLN20XW, Olympus, NA = 1). The chamber is maintained at 28.5 °C using a closed-loop temperature controller. Positioning of the chamber under the detection objective is controlled by three micro-positioning stages (M-521.DD1, M-511.DD1, and M-501.1DG, Physik Instrumente) and monitored with an infrared-sensitive camera (UI-1240SE-NIR-GL, iDS). Fluorescence emission is collected by the detection objective (located above the larva) and relayed to an sCMOS camera (ORCA-Flash 4.0 V3, Hamamatsu) through two lenses (300 mm and 200 mm focal length, respectively) and a dry objective (UPLFLN10X2, Olympus, NA = 0.3) providing a total magnification of 3×. A fluorescence filter wheel and an electrically tunable lens (ETL, EL-16-40-TC-VIS-5D-C, Optotune) are placed at the pupil of the camera objective. The ETL enables remote scanning of the focal plane in sync with the light-sheet motion, eliminating the need to move the bulky detection objective and preventing water vibrations that could disrupt volumetric imaging. Both the excitation and the detection objectives are mounted on objective scanners (PIFOC P-725.4CD, Physik Instrumente). Prior to each acquisition, the axial positions of the illumination objectives are adjusted to ensure uniform illumination of the sample. A green fluorescence filter (FF01-510/84-25 nm BrightLine, Semrock) is used to isolate GCaMP6s emission.

Collectively, the technical advancements in the setup enable functional imaging of the larval zebrafish brain with single-neuron resolution at volumetric acquisition rates of up to 5 Hz (Fig. 1b).

3 The Animal Model: Zebrafish Larvae

To perform functional imaging, we use transgenic zebrafish larvae expressing a fluorescent calcium indicator in their nervous system. Specifically, we selected the transgenic zebrafish line Tg(elavl3: H2B-GCaMP6s) [55], which carries the gene of calcium indicator GCaMP6s under the control of the elavl3 promoter, driving expression of the reporter in all neurons. In this line, GCaMP6s is fused with histone protein H2B, resulting in its nuclear localization, which is preferable to its cytoplasmic variant. This is because the signal originates from a smaller, nearly spherical structure (the nucleus), which enhances image contrast (particularly critical for 2P LSFM imaging) and simplifies neuron segmentation. The larvae used in the experiments have an albino phenotype, which, due to the absence of skin melanin pigments, ensures uniform penetration of excitation light and prevents absorption of the emitted fluorescence.

Following standard procedures [56], adult zebrafish are maintained in a freshwater aquarium system at 28.5 °C with 14/10-h light/dark artificial illumination cycle and are daily fed with both dry food and brine shrimps (*Artemia salina*) nauplii. Zebrafish larvae used in the experiments are kept at 28.5 °C in plastic Petri dishes containing fish water (150 mg/L Instant Ocean, 6.9 mg/L NaH2PO4, 12.5 mg/L Na2HPO4, 1 mg/L methylene blue; conductivity 300 μS/cm, pH 7.2) under the same light/dark cycle as the adult fish.

4 Zebrafish Larvae Preparation

Preparation for functional imaging experiments involves selecting larvae that express the fluorescent calcium indicator and employing a mounting procedure that ensures both sample stability during measurements and maintenance of physiological conditions for larval viability. Fluorescent screening is typically carried out using a stereomicroscope equipped with a fluorescence excitation kit. For GCaMP6s excitation, we use a blue LED (M470L3, Throlabs) and a pair of fluorescence bandpass filters (FF01-510/84-25, Thorlabs) in the microscope eyepieces to block excitation light. Larvae are briefly anesthetized in tricaine solution (160 mg/L in fish water; A5040, Sigma-Aldrich) so as to reduce movements, allowing us to select those with a strong green fluorescence signal in the central nervous system (Fig. 2a). The screening procedure is usually completed 1 day prior to measurements, giving larvae ample time to recover from the anesthetic.

The mounting procedure consists of the embedding of zebrafish larva in a low-concentration water-based agarose gel, providing sufficient constraint to prevent movement [5, 58]. Each larva is transferred into a reaction tube containing 1.5% low gelling point agarose (A9414, Sigma-Aldrich) in fish water, maintained in a fluid state at 38 °C. Low gelling point agarose is employed instead of standard agarose exactly because of its lower polymerization temperature, which avoids subjecting the larvae to thermal shock. The larva is then placed on a microscope slide in a drop of melted agarose, and before the gel solidifies, it is carefully positioned using fine pipette tips. This step is critical, as the orientation of the larva within the agarose determines the alignment of the light-sheet optical sectioning. For geometrically aligned transversal sections of the brain, the larva should be mounted dorsal side up, avoiding any significant tilting (Fig. 2b, top). Alternatively, the zebrafish larva can be aspirated into a thin glass capillary along with melted agarose [59, 60]. Once the agarose inside the capillary solidifies, the gel cylinder containing the larva is partially extruded, allowing fine control of the larval orientation by rotating the capillary before placing it on a microscope slide. A drop of melted

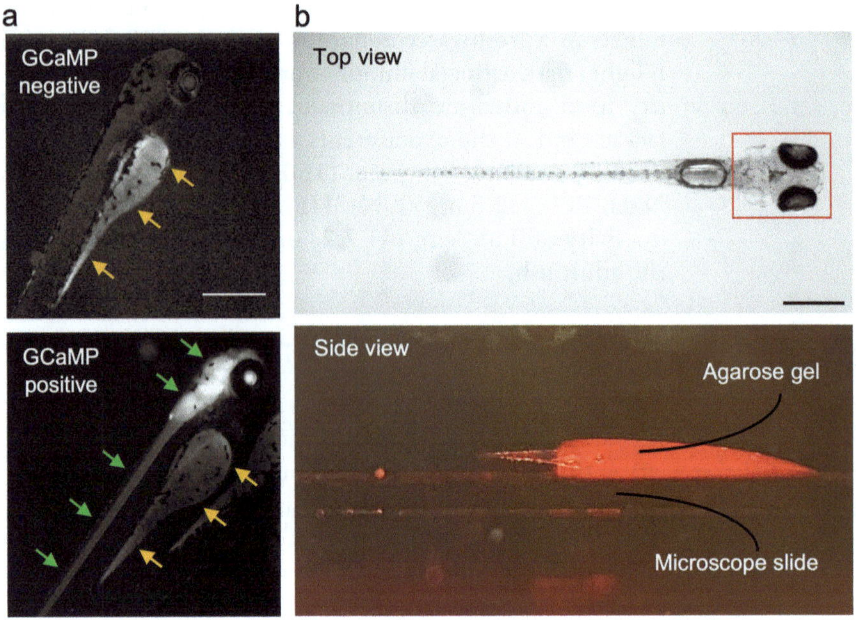

Fig. 2 Zebrafish larvae preparation: from fluorescence screening to mounting. (**a**) Representative images from the fluorescence screening procedure. Only larvae exhibiting GCaMP6s fluorescence in the central nervous system are selected (GCaMP-positive, green arrows), while non-expressing larvae are excluded (top panel). Autofluorescence in the yolk (yellow arrows) is visible in all larvae to varying degrees. Images are displayed with different greyscale levels to ensure visibility of the negative larva without saturation in the positive one. Scale bar: 500 μm. (**b**) Top and side views of an agarose-mounted larval zebrafish. (*Top*) The larva is mounted with a precisely planar orientation. The red rectangle over the head indicates the field of view acquired with 2P-LSFM. Scale bar: 500 μm. (*Bottom*) The side view highlights the microscope slide used for support and the agarose gel drop securing the larva in place (a red light from a projector was used to enhance gel visibility in water). In this preparation, agarose surrounding the tail was carefully removed with a scalpel, enabling simultaneous brain functional imaging and tail locomotor activity. (Images in panel **a** were adapted from [57], distributed under Creative Commons Attribution License (CC BY 4.0))

agarose is then applied to secure the desired position. In both methods, the polymerized agarose provides stability during imaging, while its high water content (98.5%) ensures adequate oxygen supply. To prevent any twitching during imaging, the larvae are paralyzed with a 10-min bath in 2 mM d-tubocurarine (93750, Sigma-Aldrich), a neuromuscular blocker. The microscope slide with the mounted larva is then placed in the imaging chamber filled with fish water, thermostated at 28.5 °C (Fig. 2b, bottom). Measurements are initiated immediately as the effect of d-tubocurarine lasts approximately 1.5–2 h.

5 Data Analysis Challenges

Modern microscopy often presents new challenges related to big data handling/analysis, and volumetric functional imaging with 2P LSFM is no exception. The need to capture fast biological events, such as calcium dynamics, across multiple planes encompassing the entire brain in near real-time means that the imaging system must acquire data at rapid rates without compromising on spatial resolution. This results in the generation of massive amounts of data, which need to be continuously streamed, processed, and stored.

The primary challenge lies in managing the sheer volume of data in a way that allows for efficient real-time transfer. Considering that an uncompressed 1024 × 1024 pixels 16-bit depth image is 2 MB, 1 min of 5-Hz volumetric brain imaging (31 z-planes × 5 volumes per second) generates 18.6 GB of data. Over the course of an hour, this scales up to approximately 1.1 TB. The number of individual images is also substantial: 155 per second, 9300 per minute, and 558,000 per hour. As a result, workstations must be equipped with large storage capabilities (in the tens of terabyte range) and high-throughput data streaming capabilities to prevent bottlenecks during acquisition. The use of solid-state disks (SSD), frequently organized in a RAID 0 configuration, is essential in order to achieve the necessary data-saving speeds. In this configuration, this peculiar usage leads to a noticeable decline in SSD writing performance over time, which can be mitigated through frequent memory trimming.

The acquired data require preprocessing steps before the analysis. Snapshots of individual planes are typically saved in TIFF format. Prior to analysis, these 2D images must be assembled into a 4D hyperstack, a multidimensional dataset that can be navigated through both spatial dimensions (typically representing the different depths of the sample) and in time.

As anticipated, the mounting procedure influences the position of the larval brain within the field of view. Even with the skill of an experienced researcher, manually mounted samples often exhibit slight residual tilting. During preprocessing, yaw can be easily corrected by simply rotating the images within the plane. However, roll and pitch, which affect the alignment of the optical sectioning, require more complex transformations involving rotations across multiple planes. To address this, researchers typically use open-source software (such as CMTK [61] and ANTs [62]) that enables both rigid and non-rigid transformations (morphing) to align the data with a reference brain. This reference can either be sourced from zebrafish brain atlases [63, 64] or be generated internally within the lab. Using a reference from zebrafish brain atlases offers the advantage of standardizing data alignment across labs and provides anatomical segmentation within the reference space, improving consistency and accuracy across studies.

Functional imaging data analysis is one of the most computationally demanding steps in the workflow, requiring high-performance workstations. These workstations must be equipped with substantial RAM (hundreds of gigabytes) to load entire datasets and handle the complex calculations, which often involve 32-bit file processing. Additionally, multicore processors are essential for efficiently managing the vast datasets, often containing millions of voxels per time point. Alternatively, cluster- or cloud computing can be used to process the data [65], though this approach typically requires advanced technical expertise for setup and management. Advanced computational tools, including custom scripts increasingly utilizing machine learning algorithms, are essential for automating and streamlining data processing. These tools enable tasks such as image segmentation, neuron identification, automatic detection of activation events, and eventually extraction of meaningful biological signals.

Another challenge lies in the long-term storage and accessibility of these massive datasets. Laboratories require scalable data storage solutions (such as network-attached storage, NAS) and cloud-based platforms that can not only archive this information but also make it easily retrievable for future analysis and collaborations. Storage capacity needs to be enough to store both raw data and the intermediate steps of the analysis pipeline.

6 Investigating Epileptic Seizure in Larval Zebrafish Using 2P LSFM

We employed our custom 2P LSF microscope to study the onset and propagation of epileptic seizures in zebrafish larvae. Epilepsy is a prevalent neurological disorder characterized by recurrent, unprovoked episodes of hypersynchronous and paroxysmal neuronal activity, known as seizures [66]. Despite the diverse etiological factors contributing to this condition [67], the underlying mechanism of seizures generally involves a disruption in the balance between excitatory and inhibitory neural communication in the brain [68]. Studying an animal model in such a sensitive pathological state requires minimally invasive imaging techniques. As discussed in the Introduction, the use of pulsed IR excitation, which is scantly detected by the zebrafish visual system [37], allows for gentler imaging. Compared to 1P (visible) LSFM, this approach avoids stimulating retino-recipient neurons across the larval brain, which could otherwise trigger seizure episodes in a model with a fragile balance between neural excitation and inhibition.

Seizures in zebrafish larvae are induced by bath exposure to the commonly used convulsant pentylenetetrazole (PTZ) [69], a $GABA_A$ receptor antagonist that reduces inhibitory brain activity, triggering concentration-dependent paroxysmal events resembling epileptic seizures [70]. Lower PTZ concentrations (1–2.5 mM) induce only neural hyperactivity, while higher concentrations (7.5 and saturating 15 mM) result in overt seizures [58]. Through noninvasive 2P LSFM imaging over 1-h drug exposure, we observed that increasing PTZ concentrations accelerate the rise in neuronal activity frequency (calcium peaks/min) [71] (Fig. 3a). Leveraging LSFM's optical sectioning, which enables whole-brain analysis, we evaluated the drug's effects across different brain regions. Our findings show that midbrain regions—specifically the dorsal thalamus, optic tectum, cerebellum, medial tegmentum, and interpeduncular nucleus—are particularly susceptible to PTZ, exhibiting the earliest and most pronounced increases in activity frequency [71] (Fig. 3a). Moreover, we observed distinct transitions in brain activity during seizure onset, moving through control, pre-ictal, ictal, and post-ictal regimes [71] (Fig. 3b). These different phases are characterized by unique kinetic properties and variations in the frequency of calcium transients, which can serve as a signature for each specific regime [71] (Fig. 3c). Additionally, in the pre-ictal phase, preceding overt seizures, the activities of midbrain regions display heightened correlation compared to physiological conditions, suggesting that the midbrain may function as an epileptogenic hub [71] (Fig. 3d). The study of seizure propagation across the larval brain revealed a distinctive phenomenon. While the initiation of ictal activity appears synchronous across brain regions—possibly due to activation delays being shorter than the temporal resolution of our system, i.e., 200 ms—the peaking of the activity follows a wave-like pattern. This wave consistently begins in the most caudal regions and spreads to the rostral ones. This unique propagation mechanism has led to the designation of these ictal episodes as caudo-rostral ictal waves (CRIWs) [21] (Fig. 3e).

7 Conclusions and Future Perspectives

In conclusion, the integration of two-photon excitation with light-sheet microscopy offers significant advantages for high-resolution functional imaging of biological tissues. By combining the deep tissue penetration and reduced vertebrate perception of IR illumination with the rapid, volumetric imaging capabilities of LSFM, researchers can achieve detailed visualization of complex structures with minimal invasiveness. This technique enables long-term imaging of living specimens, making it a powerful tool for studying dynamic biological processes in vivo (such as communications between neurons in neuronal networks [72]) while reducing bias from unwanted visual stimulation.

When applied to sensitive pathological models such as the zebrafish PTZ convulsive model, 2P-LSFM demonstrates its ability

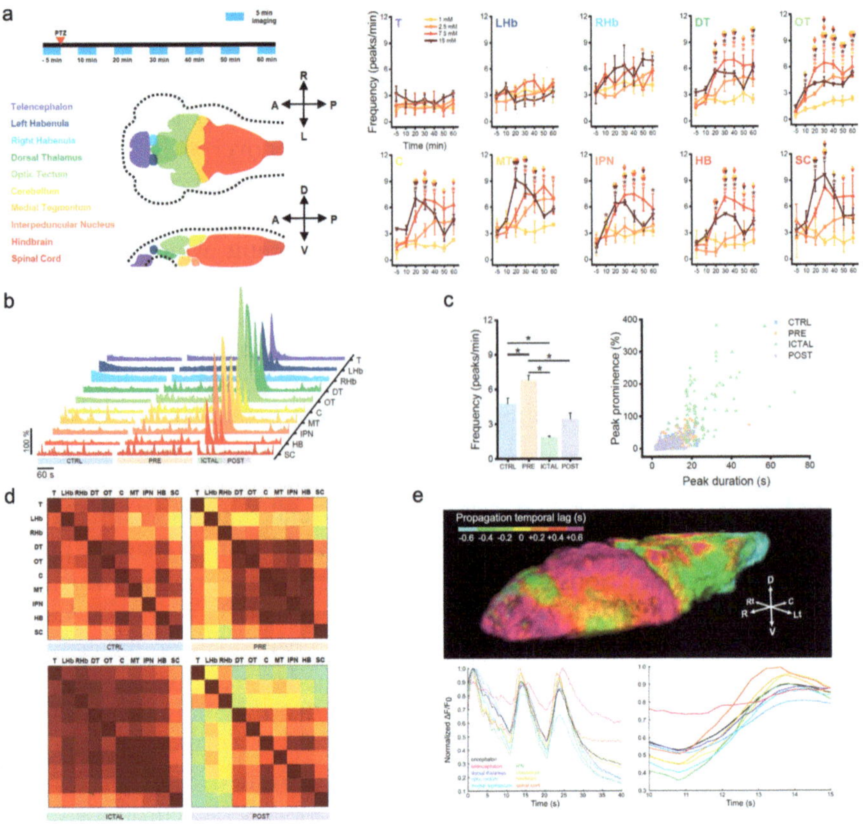

Fig. 3 Investigation of epileptic seizure in larval zebrafish using 2P LSFM. (**a**) (*Upper left*) A schematic presentation of the whole-brain functional imaging protocol. Following PTZ administration, larval zebrafish were imaged 5 min every 10 min over a 60-min period. (*Lower left*) Schematic of the larval zebrafish brain, with the ten brain regions analyzed highlighted in different colors. (*Right*) Frequency of activity peaks across brain regions (color coded as in the left schematic) at varying PTZ concentrations (1.0, 2.5, 7.5, and 15 mM, see legend for colors) during the 60-min imaging session. Midbrain regions show the earliest and most pronounced effects in response to PTZ. (**b**) $\Delta F/F_0$ traces showing brain region activity (colored as in panel **a**) during the four distinct activity regimes observed at the highest PTZ concentration (15 mM): CTRL (pre-exposure), PRE (pre-ictal), ICTAL, and POST (postictal). (**c**) (*Left*) Average frequency of activity peaks during the four phases shown in panel **b**, with each phase displaying a characteristic activation frequency. *$P < 0.05$, one-way repeated measures ANOVA with post-hoc Tukey's test. Data represent mean ± sem for $n = 3$ larvae per concentration. (*Right*) Scatterplot showing peak duration as function of peak prominence for each of the four activity regimes. Each phase exhibits distinct kinetic features in its calcium transients. (**d**) Correlation matrices displaying average pairwise Pearson's correlation coefficients of neuronal activity across the ten brain regions during the four activity regimes. Midbrain regions show increased correlation in the pre-ictal phase (PRE), which extends to the entire brain during the ictal phase (ICTAL). (**e**) (*Upper panel*) 3D rendering of a caudo-rostral ictal wave (CRIW) displaying the temporal delay of activity peaks during ictal events (see legend for colors). Caudal regions reach their activity peak earlier (negative lag values) than rostral regions (positive lag values). R: rostral, C: caudal, D: dorsal, V: ventral, Rt: right, Lt: left. (*Lower panel*) $\Delta F/F_0$ traces showing the temporal delay in activity peaks across brain regions during a single ictal event. Signals were normalized by dividing by their maximum value, with region colors corresponding to the schematic in panel **a**. Right panel shows a zoomed-in portion of the plot on the left. (Panels **a–d** were taken from [71], distributed under Creative Commons Attribution License (CC BY 4.0). Panel **e** was adapted with permission from [21], © The Optical Society)

to noninvasively access the entire vertebrate brain, enabling detailed studies of seizure onset and propagation [73]. This approach has identified midbrain regions as the areas most susceptible to the convulsant, while also revealing previously unreported caudo-rostral propagation waves across the larval brain.

Further advancements are needed to engineer solutions that enhance the signal-to-noise ratio to ultimately improve the temporal resolution of 2P LSFM. This would guarantee optimal temporal volumetric sampling, even when using the latest generation of genetically encoded fluorescent calcium indicators [74]. Additionally, incorporating a light-targeting system for 2P 3D photostimulation, using spatial light modulators [75] or acousto-optic deflectors [76], enables the creation of a fully nonlinear microscope for simultaneous whole-brain imaging and optogenetic stimulation [22], significantly expanding potential applications.

Conflict of Interest

The authors have no conflicts of interest to declare that are relevant to the content of this chapter.

References

1. Huisken J, Swoger J, Del Bene F, Wittbrodt J, Stelzer EH (2004) Optical sectioning deep inside live embryos by selective plane illumination microscopy. Science 305:1007–1009. https://doi.org/10.1126/science.1100035
2. Glaser AK et al (2022) A hybrid open-top light-sheet microscope for versatile multi-scale imaging of cleared tissues. Nat Methods 19: 613–619. https://doi.org/10.1038/s41592-022-01468-5
3. Costantini I et al (2023) A cellular resolution atlas of Broca's area. Sci Adv 9:eadg3844. https://doi.org/10.1126/sciadv.adg3844
4. Franceschini A et al (2023) Brain-wide neuron quantification toolkit reveals strong sexual dimorphism in the evolution of fear memory. Cell Rep 42:112908. https://doi.org/10.1016/j.celrep.2023.112908
5. Ahrens MB, Orger MB, Robson DN, Li JM, Keller PJ (2013) Whole-brain functional imaging at cellular resolution using light-sheet microscopy. Nat Methods 10:413–420. https://doi.org/10.1038/nmeth.2434
6. Panier T et al (2013) Fast functional imaging of multiple brain regions in intact zebrafish larvae using selective plane illumination microscopy. Front Neural Circuits 7:65. https://doi.org/10.3389/fncir.2013.00065
7. Dunn TW et al (2016) Brain-wide mapping of neural activity controlling zebrafish exploratory locomotion. elife 5:e12741. https://doi.org/10.7554/eLife.12741
8. Migault G et al (2018) Whole-brain calcium imaging during physiological vestibular stimulation in larval zebrafish. Curr Biol 28:3723–3735.e3726. https://doi.org/10.1016/j.cub.2018.10.017
9. Vladimirov N et al (2018) Brain-wide circuit interrogation at the cellular level guided by online analysis of neuronal function. Nat Methods 15:1117–1125. https://doi.org/10.1038/s41592-018-0221-x
10. Vanwalleghem G, Schuster K, Taylor MA, Favre-Bulle IA, Scott EK (2020) Brain-wide mapping of water flow perception in zebrafish. J Neurosci 40:4130–4144. https://doi.org/10.1523/JNEUROSCI.0049-20.2020
11. Göppert-Mayer M (1931) Uber elementarakte mit zwei quantensprüngen. Ann Phys 9:273–294
12. Denk W, Strickler JH, Webb WW (1990) Two-photon laser scanning fluorescence microscopy. Science 248:73–76
13. Palero J, Santos SI, Artigas D, Loza-Alvarez P (2010) A simple scanless two-photon fluorescence microscope using selective plane

illumination. Opt Express 18:8491–8498. https://doi.org/10.1364/OE.18.008491

14. Truong TV, Supatto W, Koos DS, Choi JM, Fraser SE (2011) Deep and fast live imaging with two-photon scanned light-sheet microscopy. Nat Methods 8:757–760. https://doi.org/10.1038/nmeth.1652
15. Wolf S et al (2015) Whole-brain functional imaging with two-photon light-sheet microscopy. Nat Methods 12:379–380. https://doi.org/10.1038/nmeth.3371
16. Maruyama A et al (2014) Wide field intravital imaging by two-photon-excitation digital-scanned light-sheet microscopy (2p-DSLM) with a high-pulse energy laser. Biomed Opt Express 5:3311–3325. https://doi.org/10.1364/BOE.5.003311
17. Wolf S et al (2017) Sensorimotor computation underlying phototaxis in zebrafish. Nat Commun 8:651. https://doi.org/10.1038/s41467-017-00310-3
18. Truong TV et al (2020) High-contrast, synchronous volumetric imaging with selective volume illumination microscopy. Commun Biol 3:74. https://doi.org/10.1038/s42003-020-0787-6
19. Keomanee-Dizon K, Fraser SE, Truong TV (2020) A versatile, multi-laser twin-microscope system for light-sheet imaging. Rev Sci Instrum 91:053703. https://doi.org/10.1063/1.5144487
20. de Vito G et al (2020) Two-photon light-sheet microscopy for high-speed whole-brain functional imaging of zebrafish physiology and pathology. Neurophotonics:11360. https://doi.org/10.1117/12.2560341
21. de Vito G, Turrini L, Müllenbroich C, Ricci P, Sancataldo G, Mazzamuto G, Tiso N, Sacconi L, Fanelli D, Silvestri L, Vanzi F, Pavone FS (2022) Fast whole-brain imaging of seizures in zebrafish larvae by two-photon light-sheet microscopy. Biomed Opt Express 13:1516–1536
22. Turrini L et al (2024) Two-photon all-optical neurophysiology for the dissection of larval zebrafish brain functional and effective connectivity. Commun Biol 7:1261. https://doi.org/10.1038/s42003-024-06731-3
23. Olarte OE et al (2012) Image formation by linear and nonlinear digital scanned light-sheet fluorescence microscopy with Gaussian and Bessel beam profiles. Biomed Opt Express 3:1492–1505. https://doi.org/10.1364/BOE.3.001492
24. Zhao M et al (2014) Cellular imaging of deep organ using two-photon Bessel light-sheet nonlinear structured illumination microscopy. Biomed Opt Express 5:1296–1308. https://doi.org/10.1364/BOE.5.001296
25. Takanezawa S, Saitou T, Imamura T (2021) Wide field light-sheet microscopy with lens-axicon controlled two-photon Bessel beam illumination. Nat Commun 12:2979. https://doi.org/10.1038/s41467-021-23249-y
26. Piksarv P et al (2017) Integrated single- and two-photon light sheet microscopy using accelerating beams. Sci Rep 7:1435. https://doi.org/10.1038/s41598-017-01543-4
27. Tomer R, Khairy K, Amat F, Keller PJ (2012) Quantitative high-speed imaging of entire developing embryos with simultaneous multiview light-sheet microscopy. Nat Methods 9: 755–763. https://doi.org/10.1038/nmeth.2062
28. Mahou P, Vermot J, Beaurepaire E, Supatto W (2014) Multicolor two-photon light-sheet microscopy. Nat Methods 11:600–601. https://doi.org/10.1038/nmeth.2963
29. Lin PY, Hwang SL, Lee CH, Chen BC (2021) Two-photon scanned light sheet fluorescence microscopy with axicon imaging for fast volumetric imaging. J Biomed Opt 26. https://doi.org/10.1117/1.JBO.26.11.116503
30. Saitou T, Imamura T (2023) Extended depth of focus two-photon light-sheet microscopy for in vivo fluorescence imaging of large multicellular organisms at cellular resolution. Int J Mol Sci 24. https://doi.org/10.3390/ijms241210186
31. Cella Zanacchi F, Lavagnino Z, Faretta M, Furia L, Diaspro A (2013) Light-sheet confined super-resolution using two-photon photoactivation. PLoS One 8:e67667. https://doi.org/10.1371/journal.pone.0067667
32. Chen X et al (2016) Two-photon light-sheet nanoscopy by fluorescence fluctuation correlation analysis. Nanoscale 8:9982–9987. https://doi.org/10.1039/c6nr00324a
33. Young AT (1981) Rayleigh scattering. Appl Opt 20:533–535. https://doi.org/10.1364/AO.20.000533
34. Lavagnino Z, Zanacchi FC, Ronzitti E, Diaspro A (2013) Two-photon excitation selective plane illumination microscopy (2PE-SPIM) of highly scattering samples: characterization and application. Opt Express 21:5998–6008. https://doi.org/10.1364/OE.21.005998
35. Lavagnino Z et al (2016) 4D (x-y-z-t) imaging of thick biological samples by means of two-photon inverted selective plane illumination microscopy (2PE-iSPIM). Sci Rep 6: 23923. https://doi.org/10.1038/srep23923

36. Lewis PR (1955) A theoretical interpretation of spectral sensitivity curves at long wavelengths. J Physiol 130:45–52. https://doi.org/10.1113/jphysiol.1955.sp005391
37. Jacobs GH (2012) The evolution of vertebrate color vision. Adv Exp Med Biol 739:156–172. https://doi.org/10.1007/978-1-4614-1704-0_10
38. Lemon WC et al (2015) Whole-central nervous system functional imaging in larval Drosophila. Nat Commun 6:7924. https://doi.org/10.1038/ncomms8924
39. Lee DA et al (2017) Genetic and neuronal regulation of sleep by neuropeptide VF. elife 6. https://doi.org/10.7554/eLife.25727
40. de Vito G et al (2020) Two-photon high-speed light-sheet volumetric imaging of brain activity during sleep in zebrafish larvae. Neural Imaging Sensing 2020:11226. https://doi.org/10.1117/12.2542285
41. Messina A et al (2022) Quantity as a fish views it: behavior and neurobiology. Front Neuroanat 16:943504. https://doi.org/10.3389/fnana.2022.943504
42. Luu P et al (2024) Neural basis of number sense in larval zebrafish. bioRxiv. https://doi.org/10.1101/2024.08.30.610552
43. Vladimirov N et al (2014) Light-sheet functional imaging in fictively behaving zebrafish. Nat Methods 11:883–884. https://doi.org/10.1038/nmeth.3040
44. Ricci P et al (2022) Removing striping artifacts in light-sheet fluorescence microscopy: a review. Prog Biophys Mol Biol 168:52–65. https://doi.org/10.1016/j.pbiomolbio.2021.07.003
45. Zipfel WR, Williams RM, Webb WW (2003) Nonlinear magic: multiphoton microscopy in the biosciences. Nat Biotechnol 21:1369–1377. https://doi.org/10.1038/nbt899
46. Power RM, Huisken J (2017) A guide to light-sheet fluorescence microscopy for multiscale imaging. Nat Methods 14:360–373. https://doi.org/10.1038/nmeth.4224
47. Maioli V et al (2020) Fast in vivo multiphoton light-sheet microscopy with optimal pulse frequency. Biomed Opt Express 11:6012–6026. https://doi.org/10.1364/BOE.400113
48. Gasparoli FM, Escobet-Montalbán A, Early J, Bruce GD, Dholakia K (2020) Is laser repetition rate important for two-photon light sheet microscopy? OSA Contin 3:2935–2942. https://doi.org/10.1364/Osac.405369
49. de Vito G et al (2020) Effects of excitation light polarization on fluorescence emission in two-photon light-sheet microscopy. Biomed Opt Express 11:4651–4665. https://doi.org/10.1364/BOE.396388
50. Micu I, Brideau C, Lu L, Stys PK (2017) Effects of laser polarization on responses of the fluorescent Ca(2+) indicator X-Rhod-1 in neurons and myelin. Neurophotonics 4:025002. https://doi.org/10.1117/1.NPh.4.2.025002
51. Pulin M et al (2022) Orthogonally-polarized excitation for improved two-photon and second-harmonic-generation microscopy, applied to neurotransmitter imaging with GPCR-based sensors. Biomed Opt Express 13:777–790. https://doi.org/10.1364/BOE.448760
52. Song YH et al (2025) Fluorescence spatial anisotropy of emission dipoles in an orthogonal imaging system. Opt Lasers Eng 184. https://doi.org/10.1016/j.optlaseng.2024.108649
53. Ye H et al (2022) Polarization effects on the fluorescence emission of zebrafish neurons using light-sheet microscopy. Biomed Opt Express 13:6733–6744. https://doi.org/10.1364/BOE.474588
54. Fahrbach FO, Voigt FF, Schmid B, Helmchen F, Huisken J (2013) Rapid 3D light-sheet microscopy with a tunable lens. Opt Express 21:21010–21026. https://doi.org/10.1364/OE.21.021010
55. Mullenbroich MC et al (2018) Bessel beam illumination reduces random and systematic errors in quantitative functional studies using light-sheet microscopy. Front Cell Neurosci 12:315. https://doi.org/10.3389/fncel.2018.00315
56. Westerfield M (2000) The zebrafish book. A guide for the laboratory use of zebrafish (*Danio rerio*), 4th edn. University of Oregon Press, Eugene
57. Turrini L (2020) Development of optical methods for real-time whole-brain functional imaging of zebrafish neuronal activity. Firenze University Press (FUP)
58. Turrini L et al (2017) Optical mapping of neuronal activity during seizures in zebrafish. Sci Rep 7:3025. https://doi.org/10.1038/s41598-017-03087-z
59. Sancataldo G et al (2019) Flexible multi-beam light-sheet fluorescence microscope for live

imaging without striping artifacts. Front Neuroanat 13:7. https://doi.org/10.3389/fnana.2019.00007

60. Gavryusev V et al (2019) Dual-beam confocal light-sheet microscopy via flexible acousto-optic deflector. J Biomed Opt 24:1–6. https://doi.org/10.1117/1.JBO.24.10.106504
61. Rohlfing T, Maurer CR Jr (2003) Nonrigid image registration in shared-memory multiprocessor environments with application to brains, breasts, and bees. IEEE Trans Inf Technol Biomed 7:16–25. https://doi.org/10.1109/titb.2003.808506
62. Avants BB et al (2011) A reproducible evaluation of ANTs similarity metric performance in brain image registration. NeuroImage 54:2033–2044. https://doi.org/10.1016/j.neuroimage.2010.09.025
63. Randlett O et al (2015) Whole-brain activity mapping onto a zebrafish brain atlas. Nat Methods 12:1039–1046. https://doi.org/10.1038/nmeth.3581
64. Kunst M et al (2019) A cellular-resolution atlas of the larval zebrafish brain. Neuron 103(21–38):e25. https://doi.org/10.1016/j.neuron.2019.04.034
65. Freeman J et al (2014) Mapping brain activity at scale with cluster computing. Nat Methods 11:941–950. https://doi.org/10.1038/nmeth.3041
66. Collaborators G. B. D. E (2019) Global, regional, and national burden of epilepsy, 1990–2016: a systematic analysis for the Global Burden of Disease Study 2016. Lancet Neurol 18:357–375. https://doi.org/10.1016/S1474-4422(18)30454-X
67. Holmes GL, Noebels JL (2016) The epilepsy spectrum: targeting future research challenges. Cold Spring Harb Perspect Med 6. https://doi.org/10.1101/cshperspect.a028043
68. Noebels JL (2003) The biology of epilepsy genes. Annu Rev Neurosci 26:599–625. https://doi.org/10.1146/annurev.neuro.26.010302.081210
69. Baraban SC, Taylor MR, Castro PA, Baier H (2005) Pentylenetetrazole induced changes in zebrafish behavior, neural activity and c-fos expression. Neuroscience 131:759–768. https://doi.org/10.1016/j.neuroscience.2004.11.031
70. Hansen SL, Sperling BB, Sanchez C (2004) Anticonvulsant and antiepileptogenic effects of GABAA receptor ligands in pentylenetetrazole-kindled mice. Prog Neuro-Psychopharmacol Biol Psychiatry 28:105–113. https://doi.org/10.1016/j.pnpbp.2003.09.026
71. Turrini L et al (2022) Multimodal characterization of seizures in zebrafish larvae. Biomedicine 10:951. https://doi.org/10.3390/biomedicines10050951
72. Chicchi L et al (2021) Reconstruction scheme for excitatory and inhibitory dynamics with quenched disorder: application to zebrafish imaging. J Comput Neurosci 49:159–174. https://doi.org/10.1007/s10827-020-00774-1
73. Turrini L, Roschi L, de Vito G, Pavone FS, Vanzi F (2023) Imaging approaches to investigate pathophysiological mechanisms of brain disease in zebrafish. Int J Mol Sci 24. https://doi.org/10.3390/ijms24129833
74. Zhang Y et al (2023) Fast and sensitive GCaMP calcium indicators for imaging neural populations. Nature 615:884–891. https://doi.org/10.1038/s41586-023-05828-9
75. Faini G et al (2023) Ultrafast light targeting for high-throughput precise control of neuronal networks. Nat Commun 14:1888. https://doi.org/10.1038/s41467-023-37416-w
76. Ricci P et al (2022) Power-effective scanning with AODs for 3D optogenic applications. J Biophotonics:e202100256. https://doi.org/10.1002/jbio.202100256

INDEX

Ludovico Silvestri (ed.), *Light Sheet Microscopy*, Neuromethods, vol. 229, https://doi.org/10.1007/978-1-0716-5100-1,

W

Z

MIX
Papier aus verantwortungsvollen Quellen
Paper from responsible sources
FSC® C105338

If you have any concerns about our products,
you can contact us on
ProductSafety@springernature.com

In case Publisher is established outside the EU,
the EU authorized representative is:
Springer Nature Customer Service Center GmbH
Europaplatz 3, 69115 Heidelberg, Germany

Printed by Libri Plureos GmbH
in Hamburg, Germany